The
Transcendental Part
of
Chemistry

The
Transcendental Part
of
Chemistry

DAVID M KNIGHT
Senior Lecturer in History of Science
University of Durham

DAWSON

First published 1978

© David Knight 1978

Wm Dawson & Son Ltd, Cannon House
Folkestone, Kent, England

British Library Cataloguing in Publication Data

Knight, David Marcus
    The transcendental part of chemistry.
    1. Chemistry – History – Great Britain
    I. Title
    540'.941   QD18.G7

    ISBN 0–7129–0823–4

Printed in Great Britain by W & J Mackay Ltd, Chatham

Contents

Hypotheses, treated as mere poetic fancies in one
age, scouted as scientific absurdities in the next —
preparatory only to their being altogether forgotten
— have often, when least expected, received
confirmation from indirect channels, and, at length,
become finally adopted as tenets, deducible from the
sober exercise of induction.

M. Faraday, <u>Lectures on the non-
metallic elements,</u>
J.Scoffern, (ed.)
1853, p.23.

Introduction

This is an essay on chemistry and its connection
with world views in the nineteenth century, chiefly
in Britain; where we find throughout the whole period
a surprisingly close and shifting contact between
chemical theories and philosophical positions, often
explicitly presented. The great problem for chemists
was the nature of matter. Some, but relatively few
until the last quarter of the century, opted for an
atomic theory. The standard chemical atomic theory
was that of Dalton; who posited billiard-ball atoms,
different in kind for each chemical element. Many
of his contemporaries were unhappy with these atoms,
which seemed to explain little except the definite
composition of compounds. How the juxtaposition of
oxygen and hydrogen atoms could produce water,
differing exceedingly in its properties from these
constituents, was a mystery; and it was not until
after the middle of the century that chemists,
following the hypothetico-deductive method proposed
by Laurent, were able to test propositions about the
structure of molecules.[1]

    Dalton's contemporaries at the outset of the
nineteenth century had recognised that an atomic
theory must be transcendental; that is, that if it

be assumed, then the laws of chemistry will follow deductively from it — but that the laws do not entail an atomic or any other theory of matter. That is why this essay is called The Transcendental Part of Chemistry. It was not until the beginning of the twentieth century, with the work particularly of Perrin and Einstein on Brownian movement, that independent lines of reasoning began to converge upon atoms, making it implausible to doubt their real existence. At the same time, the complex nature of the atom was revealed in the researches of J.J. Thomson on cathode rays, and of Mme. Curie and of Rutherford on radioactivity.

These last discoveries seemed to confirm what many of the greatest chemists of the nineteenth century had believed; that our simple and harmonious world was not constructed of atoms of numerous different kinds, but simply of particles of matter all of the same kind. In their different arrangements, these particles — of which the more-liberal world-definers allowed that there might be two or three kinds, rather than just one — gave rise to all the different substances which we find in the world. A belief in the unity, or very restricted diversity, of matter went back to the atomists of antiquity; and was prominent in the writings of Boyle and of Newton, who still enjoyed nearly a century after his death in 1727 an immense authority in the sciences. It is the survival of this belief from the time of Boyle to that of J.J. Thomson that we shall be especially following in this book.

We shall therefore be looking at chemical theory, and also at the background beliefs characteristic of chemists during the nineteenth

century in particular. For the doctrine of the
unity of matter is not really a theory; as we shall
see, there were a number of theories, which were
testable in various degrees, which were compatible
with it. When they were falsified, as most of them
were, or at least found to be in need of considerable modification, the doctrine nevertheless
survived. To look at the metaphysical positions and
wide-ranging theories of chemists is not to see the
whole of chemistry, or even necessarily its most
important part. As well as a transcendental part,
which belongs to the realm of intellectual history,
chemistry has a practical and a social part. The
nineteenth century saw the rise of the chemical
industry, which was the first science-based
industry requiring a stream of trained men to work
in it; and it also saw a transformation of chemical
apparatus, so that chemists at the end of the
century could readily carry out processes impossible
to Dalton's contemporaries, and could do their
analyses with very much smaller quantities. It was
also during this period that chemistry emerged as
something like a profession, with its own professional
societies in Britain — the Royal Institute of
Chemistry, and the Society of Public Analysts — as
well as its learned society, the Chemical Society of
London. At the beginning of the century, every
chemist had to make his way as best he could; by the
end, there was a regular system of chemical
education, and a career structure for trained
chemists.[2] This must not be forgotten in our story,
but it is not our primary concern.

We are involved here in intellectual history.
The Royal Institution in London, where Davy and

Faraday did their work, can be seen as a centre where landowners interested in improvements gathered to exchange ideas about stock and crops, and to hear lectures about methods of tanning or about fertilisers.[3] We shall simply be concerned with it as the locus of Davy and Faraday; for this is an essay in what is sometimes called internal history of science. No single perspective gives a complete view; but it is worth concentrating on one aspect of science if we can thus make sense of various episodes by close study of texts, even though we have to neglect complementary perspectives. We shall not however look simply at chemistry. Its boundaries have never been long constant, and at the beginning of the nineteenth century for example chemistry took over electricity following Volta's discovery of the battery, and soon afterwards took over mineralogy too as a branch of inorganic chemistry and crystallography. By the end of the century, electricity had been given up to physics; and many physicists believed or hoped that with their kind of atomic theory they would soon take over the theory of matter from chemists. Ever since the days of Newton, there had been those who hoped to reduce chemistry to mechanics; while there had also been upholders of a distinct chemical philosophy, who urged the autonomy of their science and pointed to the futility of many physicists' excursions into chemistry. Some of these pointed to parallels between chemistry and natural history; and in the last quarter of the nineteenth century chemists preoccupied with classifying the elements did indeed turn their attention to Darwinian biology.

We cannot confine our attention simply to

science either, for the boundaries of science are
not sharp and the sort of background beliefs with
which we are concerned are not derived simply from
experiments. Men making voyages in strange seas of
thought, or wrestling with nature in the dark, have
to feel their way as best they can. Nor shall we be
confined to a single perspective, for this seems to
be a story that cannot be simply told chronologically.
We shall begin with two chapters to set the stage;
leading up to about 1800, when Lavoisier's
definition of a chemical element had become accepted
despite its curious negative — empirical form — for
an element is a substance that cannot be decomposed,
and therefore this year's element might next year
have to be removed from the list if somebody invents
a new technique of analysis.

In the following chapters we follow the schema
of Auguste Comte, for whom knowledge progresses as a
theological stage gives way to a metaphysical, which
in its turn gives way to a positive or really
scientific stage. This is a thesis about which the
historian ought to be completely agnostic, for he
must not impose a thesis upon his evidence; and the
Comtean schema here is simply a way of identifying
three viewpoints from which distinct but overlapping
perspectives on the doctrine of the unity of matter
can be displayed.

The first of these is to see it in relation to
materialism. Especially in Britain, the outbreak of
the French Revolution and its development into the
Reign of Terror, was attributed to the teachings of
men such as Diderot, Voltaire, and d'Holbach; who
had urged that there was nothing in the world but
matter, and no room for human souls or for free

choice. On this view, chemistry like everything else could be reduced to mechanics. Among Davy's great achievements in the early years of the nineteenth century, in his own eyes and in those of contemporaries, was to show that mechanics was not enough. Chemistry was a science of immaterial powers such as electricity as well as of matter; the same matter could form a diamond or a lump of charcoal, and it was the forces arranging the matter which determined its properties. Matter, on this analysis, was brutish and inert, and there was no real reason to suppose that there were different kinds of it.

The next perspective is to look at the fourth state of matter, a topic which features prominently in the writings on cathode rays of the great free-lance chemist William Crookes, writing in the last quarter of the nineteenth century. He got the idea that all matter might be put into a fourth state, as different from gases as gases are from liquids, from Faraday's early writings. In fact, it was a common-place in the early nineteenth century, being derived from Newton's corpuscular theory of light; for in one of the Queries at the back of his Opticks, Newton had asked whether all matter might not be converted into light, clearly expecting the answer 'yes'. Crookes absorbed from Faraday a firm belief in the unity of nature, and in the light of this metaphysical view and his doctrine of a fourth state, he interpreted the data from his cathode-ray tube and his spectroscope as indicating that the various chemical elements were all really different arrange-ments of particles of one prime matter.

The next chapters deal with the 'positive' phase of the inquiry; beginning with Davy's work on chlorine in which he demonstrated its analogies with oxygen. As more elements were discovered in the first half of the nineteenth century, it became increasingly implausible to suppose that such similar substances as sodium and potassium, or chlorine and iodine, were really quite irreducibly different. In chapter 6, we consider Prout's Hypothesis that all elements were composed of hydrogen; an attempt to move from analogical arguments to quantitative ones, for if it were true all atomic weights should be multiples of that of hydrogen. In the event, this turned out not to be true; but many were close to whole numbers, and although apparently refuted Prout's hypothesis lingered on throughout the century. In the next chapter, we see arguments derived from crystalline and molecular structures, which indicated that elements might be like the radicals of organic chemistry, stable groupings of atoms of only two or three kinds.

The eighth chapter is concerned with the coming of energetics into chemistry in the second half of the nineteenth century, and then with the chemical calculus of Benjamin Brodie, an attempt to rid chemistry of hypotheses which was in fact received as an argument for the unity of matter. Throughout these four chapters we shall meet attempts to achieve transmutations, which sometimes seemed to have been crowned with success, but which always turned out in the end to be false. The doctrine of unity of matter remained as potent as ever, despite these falsifications; for chemists were not so hard-headed as to give up hope of demonstrating the simplicity and harmony of nature.

The ninth, and last, chapter is concerned with chemical taxonomy; with the success ultimately in finding, in the Periodic Table, a system of classification of the chemical elements. Since this was achieved some ten years after the Origin of Species had come out in 1859, it was not surprising that the Table was interpreted in an evolutionary manner. Transmutations might not now be possible, but it seemed more than probable that all the elements were in some sense descended from hydrogen. The plausibility of this view became apparent with the researches of the atomic physicists around 1900, with their atomic models composed of protons and electrons. Belief in the unity of matter, held through thick and thin by chemists in the nineteenth century, seemed vindicated; and no doubt it was because of this background of thought that Rutherford's interpretation of radioactivity as transmutation found surprisingly ready reception — it was just what many chemists had been hoping for, and indeed asking for in Presidential Addresses and similar public occasions. Our story then comes to an end, for the transcendental part of chemistry has turned into atomic physics.

NOTES

1. J.H. Brooke. "Laurent, Gerhardt and the Philosophy of Chemistry", Hist.Studies Physical Sciences, VI(1975) 405-30.

2. C.A. Russell, N.G. Coley, and G.K. Roberts, Chemists by Profession (London,1977).

3. M. Berman, Social Change and Scientific Organization: the Royal Institution 1799-1844, (London,1978).

1

The latent and Less Obvious Properties
of Bodies

'The transcendental part of chemistry' is a phrase
used by Humphry Davy in a lecture of 1809; a lecture
comprising, as his brother remarked, "the author's
conjectures relative to several interesting
speculative points". The phrase, though it sounds
Kantian, may be found also in contemporary translations of Lavoisier and of the Swedish chemist
Tobern Bergman; the latter translation was made by
the down-to-earth philosopher Jeremy Bentham, and the
term does not therefore indicate any influence from
German Idealism.[1] But it does have a curious sound
to modern ears, which may serve to remind us of how
different the chemistry of this epoch was from the
'normal science' learned by the chemists of today.
The chief object of this study will be to explore
these differences. We shall therefore be more
concerned with recovering the views of chemists,
usually prominent but occasionally obscure, of the
late eighteenth and the nineteenth century rather
than with seeking in their utterances apparent
parallels with chemical doctrines now in vogue.

There are then two related points to bear in
mind. The first is that there were very few
professional chemists until the nineteenth century
was well advanced. The science was advanced by
physicians, apothecaries, manufacturers, clergymen of

various denominations, and gentlemen of independent means, as much as by such professors of chemistry as there were.[2] Such men tended to be interested in the broader aspects of their science, and anyway could not find a large audience of narrow professionals for highly technical papers. The second point is that the pursuit of knowledge for its own sake, which we tend to take for granted as the basis for theoretical science, did not seem quite respectable until remarkably recently. The Orientalist Sir William Jones, and the neoplatonist Thomas Taylor,[3] who sought to restore the worship of the ancient gods in the England of Arkwright and Watt, proclaimed themselves devotees of truth, prepared to go wherever it might lead; but in studies of nature some further motive seems usually to have been expected. One should wish to improve the lot of one's fellow men, as the founders of the Royal Institution and the Mechanics' Institutes did; or going beyond this 'vulgar utility' — as Davy called it — one could use the science as a basis for a world-view.[4] The study of nature has always had its devotees; but a science that can cast light upon nature, man, and God cannot but be of immense importance and general interest. In examining the transcendental part of chemistry, therefore, we shall not be simply resurrecting that part of the science which dealt with entities which are not directly observable, but will be seeking to view chemistry in its relations to philosophy and theology.

This may seem an unlikely light in which to view chemistry; we know that astronomy, at least from Galileo, through the controversy between Leibniz and the Newtonian Samuel Clarke, to Laplace,

raised such questions; and so did biology, most
notoriously in the Darwinian controversy. Chemists,
on the other hand, do not appear likely to provoke
denunciation, nor merit commendation from the pulpit,
nor to find their discoveries enshrined in poetry or
employed as illustrations in works of philosophy.
It was not always so; around 1800, great audiences
flocked to hear lectures on chemistry, and tickets
changed hands at inflated prices on the black
market. These crowds did not simply go to see
spectacular demonstration-experiments, though these
were undoubtedly an attraction; they went to learn
some natural — or chemical — philosophy. The
sciences of chemistry and electricity, which at this
time were closely linked through the discoveries of
Galvani and Volta, promised to disclose the inner
structure of matter and cast light upon the nature
of the mysterious powers — chemical affinity, heat,
light, electricity, magnetism, and gravity — by
which matter was arranged into the various bodies
which we see, and endowed with motion. Schelling
and Coleridge were among those who took an immense
interest in chemistry; and Goethe's novel Elective
Affinities applied to marriages the dynamical theory
which Bergman had developed to account for double-
decomposition reactions in the inorganic realm.[5]

    There were at least two features of the science
which made it particularly attractive. The first was
that it was theoretically incoherent. There was
relatively little agreement upon chemical theory,
though this made surprisingly little difference in
the practical part of the science with which the
multi-volume Systems of Chemistry of about 1800 are
distended. Unless one was a chemist, there was no

3

need to wade through such treatises on advanced
cookery; one could instead take an interest at once
in controversy. The Phlogiston theory had received
a heavy blow from the researches of Lavoisier; but
it was not dead, and there were widespread hopes and
fears until at least 1810 that Davy would restore it
and overthrow the French doctrines. The Newtonian
dynamical theory, according to which chemistry was
to be quantified as astronomy had been by measurements of forces, had led to studies on elective
affinities by Bergman, by Berthollet, and by Thomas
Young among others; but it was now showing signs of
collapse as Dalton's statical atomism began to
achieve a quantification of chemistry in terms not
of forces but of weights.[6]

One way of interpreting the history of chemistry
through the nineteenth century is as a struggle
between the dynamical Newtonian atomism and the
Daltonian chemical atomism. Berzelius' dualistic
theory of chemical combination was a first attempt
at a reconciliation; but it broke down when confronted with the data of organic chemistry. The work on
molecular structure by Kekulé and his school
represents the great triumph of Daltonian atomism;
but J.J. Thomson's paper on cathode rays, which he
interpreted as composed of corpuscles, put Newtonian
atomism back in the forefront. While Dalton
proposed irreducibly different atoms for each
chemical element, requiring therefore that the world
was composed of many different building-blocks;
Newtonians believed in one prime matter, defined in
terms of its mass and not of its chemical properties,
which was arranged into different bodies by powers
which it was the duty of natural philosophers to

4

investigate. Both sides held that their world view
was simpler; while others agreed, following
Lavoisier, that any atomic theory was bound to be a
matter of metaphysics, and that the chemist's task
was not to arrange hypothetical particles but to
discover facts and laws connecting them.

If there was considerable doubt then as to how
the facts of chemistry should be interpreted, there
was equal uncertainty as to their significance.
This brings us to the second attractive feature of
chemistry; that it promised to confirm or refute the
doctrine of materialism. From at least the time of
Hobbes, it had been feared, particularly in Britain,
that the adoption of materialism must lead to im-
morality and civil disorder, or at best to a cold
selfishness; the events in France as the Revolution
led to the Reign of Terror and the Napoleonic
tyranny served to confirm these melancholy
predictions. By 'materialism' writers of the
eighteenth century did not mean the theory denounced
from pulpits in our day, that more possessions will
bring happiness; but the doctrine that only matter
existed,[7] and that there was no such thing as
immaterial substance. Man hence had no immortal soul,
and so there was no reason why he should not make himself comfortable.

More seriously, if matter was all that existed,
it must have inherent in it the powers by which it
was moved and arranged. Newton had argued that
gravity could not be inherent in matter, in a passage
that Faraday loved to quote; and it became a necessary
part of the creed of the immaterialist to assert that
matter existed, but was utterly inert and passive.
The materialist, on the other hand, required matter to
be active; and in the materialism of Joseph Priestley,

5

the most notorious English advocate of the system,
the matter has — paradoxically — almost disappeared
into mere centres of force or power in a world which
is mostly void.[8] In chemistry, the study of forces
or powers might illuminate this area; and the study
of chemical synthesis, in which two substances
combine to form a third having very different
properties from its components, seemed especially
promising and recalled discussions of the union of
soul and body to form a person. Materialism was
moreover raised directly in discussions of chemical
properties. If a body were combustible, then
according to Becher and Stahl this was because it
contained phlogiston (φλογιστος , inflammable) which
was probably a material substance. Certainly for
Lavoisier, if a body were hot or if it were a gas, it
contained the material substance caloric; and if it
were acidic, then it contained the material substance
oxygen. In Germany and in England, both these ideas
of Lavoisier's were attacked by dynamical theorists;
and it seemed that if, for example, it could be
demonstrated that there was no material constituent
common to all acids, then materialism would have been
overthrown.

It will therefore be necessary for us, if we are
to understand the transcendental part of chemistry in
the early nineteenth century, to recover both the
specifically chemical background and also part at
least of the general intellectual background;
particularly the doctrine of substance, and the
current philosophy of science. We shall begin therefore with a discussion of the theory that the basic
entities with which the chemist should concern himself
are elements. This view had been combatted by Boyle,[9]
who is rather curiously still often credited with

penning the first adequate definition of the term
'element'; he wished chemists to take as their basis
the atoms or 'corpuscles' of matter of the
physicists. Newton and the Newtonians followed
Boyle in this, but made the determination of the
forces between the corpuscles the object of the
chemist; but Lavoisier urged chemists to accept the
limits of analysis as elementary and to forgo atomic
speculations. Dalton then made the smallest
particles of these elements his atoms; and diverted
chemistry from its pristine but sterile Newtonian
path.

Although it seems that Dalton's atomic theory
had theological foundations — in Quaker orthodoxy —
it would appear that his distinct atoms of iron,
oxygen, copper, and so on diverge not only from the
Newtonian orthodoxy but also from ordinary immaterialism, for these atoms have a range of
interesting properties whereas they should be merely
inert and massy. We must therefore look in more
detail at materialism and science in the eighteenth
and the early nineteenth century; and this will help
us to see why it was that chemistry was the science
which perhaps most interested those usually described
as Romantics. In this context we shall look at the
work of Davy, himself a Romantic and the most
glittering chemist of his era.

As the century wore on, the various forces or
powers came to be seen as forms of energy, and the
bearing of chemistry upon materialism became less
obvious. The science began also to develop theories
of increasing power, and the scope for the amateur
decreased. But the great question remained; were
the elements irreducibly different, or were they all
simply stable arrangements of the same basic

particles? As the number of known elements increased — and the Voltaic battery, the spectroscope, and apparatus for liquefying gases were responsible in turn for adding large numbers to the list — so it seemed less plausible to suppose that the simple and harmonious world in which we live could be composed of them. The success of Mendeleev in 1869 in grouping them in families suggested that like the species of the animal and vegetable kingdom, the various elements were in some sense descended from a common ancestor.

Spectroscopic observations seemed to support this conclusion; which received further support from a Newtonian survival, belief in a fourth state of matter or in the possibility that matter and light were interconvertible. For the corpuscularian, the transmutation of one element into another was perfectly conceivable; indeed, it was often supposed to occur in physiological processes. In the nineteenth century, as we shall see, various transmutations were reported, none of which could be confirmed or repeated. But this did not produce general scepticism; the study of radioactivity soon produced with the work of Ramsay and Soddy the first transmutation to be followed in the laboratory; and with the atom of Rutherford and the X-ray studies of Moseley the elements disappeared as basic entities, and the transcendental part of chemistry became atomic physics.

While we shall be chiefly concerned with the nineteenth century, the story has to begin earlier; and the proper point of departure seems to be the Sceptical Chymist of Robert Boyle. This book, which first appeared in 1661, is more often referred to than read, and to this fact probably owes its high

reputation. Its object was to undermine the current
theories of elements and to urge — in a general way
— mechanical and corpuscularian explanations instead.
There were two current theories of the elements:
that derived from Empedocles and taken up in
opposition to atomism by Aristotle, and that of
Paracelsus, which seems to have its roots in the
alchemical tradition. Aristotle's elements were
earth, air, fire, and water; they were not
immutable, for matter was a substrate in which one
each of various pairs of qualities inhered. He put
forward subtle argument for the view that two pairs
of qualities, hot/cold, and wet/dry, were fundamental;
the four possible combinations of one each of these
pairs generated the four elements. The elements
differed from the earth, air, fire, and water which
we find around us; for all terrestrial things
contained some of each element. As we shall see,
the Aristotelian view survived Boyle's criticisms,
and lasted — in a modified form — through much of
the eighteenth century.

By the 1660s, however, most chemists had
adopted the Paracelsian doctrine, which Boyle
referred to as the 'chymical' theory; for this was a
period of great interest in alchemy and magic.
According to Paracelsus, the four elements were
composed of three principles; sulphur, mercury, and
salt. He had added salt as a conservative,
incombustible principle to the sulphur and mercury
of the alchemists, writing: 'whatever burns is
sulphur, whatever is humid is mercury, and that which
is the balsam of these two is salt.'[10] Like the four
elements, the three principles were ideal entities
which could not be isolated. They simply resembled

9

in certain respects the ordinary substances after
which they were named, all of which in fact contained
all three. While an explicitly Paracelsian theory
does not appear to have lasted into the eighteenth
century, 'philosophical' alchemists persisted; a
notable example was Peter Woulfe, the inventor of a
wash-bottle for gases, who was a Fellow of the Royal
Society and lived into the nineteenth century. But
this alchemy owed at least as much to Boyle and
Newton as to Paracelsus.

The Sceptical Chymist is organised — if we may
stretch the term — in the form of a dialogue, which
was then a popular way of writing, even for works of
science.[11] The plan seems to be to demonstrate the
following: that absurdities follow from the dogma
that fire is the only, or paramount, agent for
analysis; that the products of such analysis were
not necessarily constituents of the original body;
that analyses do not always lead to three or four
components; that such products of analysis lack the
purity and simplicity which true elements should
have; and that it would be better to stop talking
about elements and adopt instead mechanical,
corpuscularian accounts of the phenomena. Boyle, in
the person of the sceptical philosopher Carneades,
insisted that all arguments must be from experiment,
not reason; a demand which seriously weakened the
Aristotalian case. Further, since he is a sceptic,
he claims the right to oppose rather than to present
a detailed programme of his own.[12]

The most serious problem for Carneades'
opponents in the debate is that what they mean by
'element' — and what Boyle also seems to have meant
— is something quite different from what we mean.

For the chemist of the seventeenth century, everything in the world was a blend of all the elements
or principles. His elements are therefore closer to
our protons and electrons, which are found in every
atom, than to our elements — oxygen, carbon, and so
on — which are not found in every substance which we
subject to analysis. It is clearly hopeless both to
believe in ideal,omnipresent elements, and also to
expect to analyse substances into their elements; and
chemists seem to have been satisfied with analogies.
Thus when a green stick burnt, the smoke indicated
'air', the flames 'fire', the ash 'earth', and the
hissing 'water'. Similarly, Paracelsians claimed to
derive from analyses a spirit, an oil, and a salt; a
caput mortuum or residue also remained, but this was
of little significance. As Boyle was not slow to
remark, this is not always what happens when inorganic
materials are heated; and even when organic materials
are distilled, the products vary widely under
different conditions. Furthermore, the oils and
salts produced from different bodies differ in taste
and in crystalline form, which they should not do if
they were true elements. The classification of
substances as spirit, oil, or salt was not even unambiguous, depending as it did upon taste,
combustibility, and solubility in water.

Having at great length established the
inadequacy of both received opinions, Boyle proceeds
to give a sketch of how the corpuscular philosophy
might in principle be applied to explain the phenomena
of chemistry. The section begins with an assault
upon the chemists for presuming to call themselves
philosophers:

> But alas, how narrow is this philosophy,
> that reaches but to some of those

compound bodies, which we find but
upon, or in the crust or outside
of our terrestrial globe, which is
itself but a point in comparison of
the vast extended universe, of whose
other and greater parts the doctrine
of the tria prima does not give us
an account![13]

Magnetism and physiology also receive no
explanation from the Paracelsians; their theory therefore neither works in chemistry, nor links chemistry
to other sciences. With the corpuscular philosophy
Boyle hoped to join chemistry securely to physics.
To him, an attraction of atomism was that it seemed
capable of explaining everything; to some critics
like Stahl it appeared that in trying to explain
everything, it explained nothing. Boyle believed in
an hierarchy of particles; 'primary mixts' were stable
configurations of corpuscles. which survived unchanged in chemical reactions in the course of which they
were rearranged. In the Sceptical Chemist, Boyle
tried to show that it was not unreasonable to
suppose that qualities such as colour might be
explained in terms of arrangements of corpuscles; and
he amplified these accounts in other works.

Within the 'Paradoxical Appendix' to the work
we find the definition of elements[14] as, 'certain
primitive and simple, or perfectly unmingled bodies;
which not being made of any other bodies, or of one
another, are the ingredients of which all those
called perfectly mixt are immediately compounded, and
into which they are ultimately resolved ...'. This
was not a new definition, and Boyle did not claim
that it was; these elements were the omnipresent
principles in which Boyle did not believe, and he

went on to give more reasons for doubting whether there are any such things. When Lavoisier revived the notion of elements, he was, in effect, removing chemistry from a basis in the atoms of natural philosophy back onto its own foundations. For Boyle and the corpuscularians, the atoms possessed only the primary qualities; if they differed at all, it could only be in size and shape. As in the eighteenth century, the theory became increasingly dynamical and the atoms became smaller and smaller until the whole matter of the solar system might be supposed to occupy no more space than a nutshell; so any difference in size or shape of the atoms, now mere points, ceased to be possible. By a rearrangement of its corpuscles, then, any substance might be converted into any other; and the conversion of, for example, base metals into gold might reasonably be hoped for.

Thus in 1811 Davy spoke of the growth of metals in a lecture:

> Even in these times of a more exalted science, after a series of connected discoveries, we have no right to say, that what Stahl suspected — the generation of gold in nature — does not take place. We, it is true, have never seen it composed or decomposed; but our works are in moments, — those of nature in ages; ... I hope I shall not be misunderstood on this delicate subject, or from what I have said be considered as an alchemical projector. ... To search for the elixir of life, or the powder of projection, would be a mark of a feeble, prejudiced, and ignorant mind. But to enquire whether the metals be capable of being decomposed and composed is a grand object of true philosophy.[15]

Davy was here in the corpuscularian tradition, as

was Faraday when he suggested that the alchemists' doctrine of transmutation was not false, but a vision of truth, distorted. Changes in chemical properties, and qualities generally, were to be explained in terms of new arrangements, or 'textures', of the corpuscles. The problem was that such explanations were post facto and very general; no predictions or detailed explanations could be given of the facts of chemistry.

It was Newton who turned the theory towards dynamics.[16] The success of his labours in the fields of mechanics and astronomy led to the hope that the history of chemistry would take a similar course; that the laws of chemical affinity would be determined by a Kepler of chemistry and then shown by a second Newton to depend upon some single law of forces. As Newton had unified celestial and terrestrial dynamics, so this second Newton would unify the macroscopic and microscopic worlds, demonstrating that the forces between particles were similar to — perhaps even identical with — those acting upon bodies large enough to be observable. Arnold Thackray has shown that this view guided the majority of chemists during the eighteenth century; and that chemistry did not suffer, as has been suggested, from a delayed scientific revolution but from what turned out to be a premature and inappropriate scientific revolution along Newtonian lines. The influence of Newton persisted long into the nineteenth century, as we shall see; particularly among those whom with slight anachronism can be described as physical chemists, interested in chemical dynamics.

It was particularly to the Queries appended to

Newton's Opticks that chemists turned;[17] we find
Davy, for instance, quoting from one of the Queries
in a lecture in the early nineteenth century. For
Newton, the corpuscles were held together by
'attractions',[18] a descriptive not explanatory term:
'There are therefore Agents in Nature able to Make
the Particles of Bodies stick together by very
strong Attractions. And it is the Business of
experimental Philosophy to find them out.' Elsewhere he proposed an hierarchy of particles, and
gave some indications of how the dissolution of
silver in aqua fortis (our nitric acid), and of gold
in aqua regia but not aqua fortis might be accounted
for in mechanical terms. Laplace and Lord Kelvin in
the nineteenth century toyed with the idea that the
same power might be responsible for gravity and for
chemical affinity; but the latter is elective whereas
the former is universal, and Newton himself invoked
'some other attractive and repelling powers which
intercede the particles.' Repulsive as well as
attractive powers were required, since solid bodies
are clearly not completely packed with matter. If
the atoms are defined in terms of mass, then clearly
a given volume of gold contains many more than an
equal volume of iron, which is lighter and hence must
include more void. Even the gold can be made to
contract by cooling, and must contain void spaces;
some repulsive power is clearly called for here.
Newton poured scorn upon the hooks which atomists of
antiquity had invoked to explain how particles
cohered; and his writings, with their dynamical
emphasis, seem to have put an end to the view that
atoms might differ in size and shape.

The atoms of the corpuscularians, then, were

atoms of matter. They thus fitted tolerably well
the demand that one makes of an atomic theory, that
it should account for complex properties in
terms of different arrangements of atoms which have
much simpler properties than the bodies. But it is
one thing to give a general account of the genesis
of physical properties, and quite another to explain
in any detail even the simplest facts of chemistry,
a science concerned with the interaction of what
appear to be different sorts of matter. Chemists in
eighteenth-century Germany followed another course;
and one can see why, even if one has some sympathy
with Partington's contrast between 'the path of
discovery opened out by Boyle, Hooke, and Mayow' and
the 'jungle of the theory of Phlogiston'.[19] As
Stahl remarked: 'Mechanical philosophy, though it
vaunts itself as capable of explaining everything
most clearly ... scratches the shell and surface of
things, and leaves the kernel untouched.' (See note
21).

In similar vein, William Whewell in the
nineteenth century wrote[20] that the corpuscular
philosophy in chemistry was pretentious and super-
ficial; he approved of a split between chemistry and
physics, believing that every science must have its
own assumptions and its own sphere, and that
confusion must result when the facts of one science
are 'explained' in terms of the postulates of another.
Mechanical principles had not been applied, he
believed, to give a definite explanation of any
particular chemical phenomenon. The assumption of
particles consisting of a congeries of smaller
particles held together by forces was no better than
the assumption of atoms having hooks; for as well as

hypothetical particles we have to postulate these
hypothetical forces. The view that chemical
affinity and gravitational attraction were analogous
had never been worked out in sufficient detail to be
of any use, and any attempt to work it out would
show its inadequacy; the phenomena of chemistry
could not be explained mechanically without an
immense accumulation of hypotheses. This drubbing
of the Newtonians reminds us of Cotes' treatment of
the Cartesians in his edition of Newton's _Principia_;
but that Whewell should in the middle of the nineteenth century find it necessary thus to attack the
corpuscularians is some indication of the continuing
vitality of Newtonian science.

It seems that Becher, the founder of the theory
of phlogiston, supposed omnipresent elements like
those of the Paracelsians; but avoided the problem
of distinguishing them from ordinary substances by
calling them _terra pinguis_, _terra fluida_, and _terra
lapidia_, instead of sulphur, mercury, and salt.[21]
His pupil Stahl introduced the term _Phlogiston_ in
place of _terra pinguis_. On grounds similar to
Boyle's, he rejected both the Aristotelian and
Paracelsian elements, pointing out that nobody had
been able to show that the 'principles' obtained
from all bodies upon analysis were the same. Though
he wrote that: 'Universal chemistry is the art of
receiving mixt, compound, or aggregate Bodies into
their Principles; and of composing _such Bodies_ from
those _Principles_', he seems to have thought that in
fact the products of analysis were 'mixts' rather
than principles.

Throughout the eighteenth century we find
speculation about the elements, as well as attempts

17

to apply the corpuscular theory in chemistry. Thus Scheele wrote[22] at the beginning of his paper describing the discovery of what we call oxygen that it was the business of chemistry, 'skilfully to separate substances into their constituents, to discover their properties, and to compound them in different ways'. He deplored any tendency among chemists to stop looking for elements, using the term to mean omnipresent principles:

> Hitherto chemical investigators are not agreed as to how many elements or fundamental materials compose all substances. In fact this is one of the most difficult problems; some indeed hold that there remains no further hope of searching out the elements of substances. Poor comfort for those who feel their greatest pleasure in the investigation of natural things! Far is he mistaken, who endeavours to confine chemistry this noble science, within such narrow bounds!

Some chemists adhered to the Aristotelian principles; others believed that air and phlogiston were such elements; and Scheele hoped in writing the <u>Treatise on Air and Fire</u> to clear the matter up. He began with the study of fire, because without fire it would be impossible to perform experiments; and then found it impossible to form a true judgment of fire without knowledge of air. So it seems that it was his concern about the elements which led Scheele to the experiments which culminated in the discovery of pure air, or oxygen.

In England, the study of gases had been initiated by the Newtonian Stephen Hales,[23] who found that many solid bodies contained 'air' which could be liberated by heat. His technique prepared the way for Priestley, Cavendish, and Lavoisier; but his

interpretation of his observations led to a revival
of belief in the four Aristotelian elements. If
most bodies contained air, they might well contain
fire, earth, and water too. Thus the French chemist
P.J. Macquer,[24] a generation older than Lavoisier,
insisted that chemistry was a simple science founded
on facts. Rather surprisingly, he believed that
experiment revealed five elements, the Aristotelian
quartet and phlogiston. In analyses, he wrote, 'We
are always stopped by substances in which we can
produce no change, which are incapable of being
resolved into one another, and which stand as so many
barriers obstructing our progress.' These may not be
ultimate elements, but they are substances the
composition of which we cannot discover. The five
elements were not omnipresent, and some substances
might lack one or more of them; thus metals were
devoid of air and water.

Davy contrasted Scheele — a faithful disciple
of Bacon and Newton, who took hypotheses with
some seriousness — with Priestley,[25] who 'paid but
little attention' to theory, forming hypotheses
rapidly and relinquishing them 'with an ardour almost
puerile'. Examining Priestley's exuberant chemical
writings, one can see Davy's point; but in Priestley's
independent discovery of what we call oxygen, which
Priestley himself modestly attributed to chance, this
readiness to abandon earlier views was very important.
As he wrote:

> There are, I believe, very few maxims
> in philosophy that have laid firmer
> hold upon the mind, than that air,
> meaning atmospherical air (free from
> various foreign matters, which were
> always supposed to be dissolved, and
> intermixed with it) is a simple

elementary substance, indestructible,
and unalterable, at least as much so
as water is supposed to be. In the
course of my inquiries, I was, however,
soon satisfied that atmospherical air
is not an unalterable thing ...

Priestley's great achievement was to move from on
the idea that the 'air' [oxygen] derived from the
calcination of mercuric oxide was 'at least as good'
as common air — a statement Hales might have made —
to the notion that it was a distinct species of air.
Henceforward, air could not be treated as an element,
any more than salt could after seventeenth-century
workers had shown that there were several different
salts. The status of water was not determined by
Priestley, and remained an open question into the
nineteenth century, as we shall see in due course.

In fact, Joseph Black[26] had shown some twenty
years before Scheele and Priestley had published
their researches that 'fixed air' [our carbon
dioxide] was different from ordinary air. But unlike them, he does not seem to have taken the
Aristotelian list of elements seriously. Black seems
to have been sceptical about the value of hypotheses
in chemistry; he thought that the sciences had been
'obstructed by speculations about the principle of
affinity, and, particularly, by the attempts of
ingenious men to explain the chemical operations by
attractions and repulsions'. To propose mechanisms
was usually a mere waste of time; at best they could
be but analogies, not true explanations. When a
corpus of information like that which Newton had
established in astronomy was available, then one
might hope to explain the laws of affinity. This
reminds us of the remarks on causes and hypotheses of
Black's contemporary, the 'common-sense' philosopher

Thomas Reid. Lecturing about the elements, Black
said: "We shall not pretend any further than can be
shown by experiment," a sentiment which Lavoisier
was to echo.

Lavoisier's ideas of the status of the elements
were not entirely new, nor did he claim that they
were; but they were very influential and seem to
have crystallised the thoughts of many of his
contemporaries. He stated clearly what many others
had felt indistinctly; and his writings signal the
end of the long reign of the Aristotelian elements.[27]
He remarked that while many writers supported the
doctrine of the four elements, they all found that
more were required to fit the facts; and he explained
their errors in condescending terms, as befitted a
product of the Age of Reason:

> All these chemists were carried
> along by the influence of the
> genius of the age in which they
> lived, which contented itself
> with assertions without proofs;
> or, at least, often admitted as
> proofs the slightest degrees of
> probability, unsupported by that
> strictly rigorous analysis required
> by modern philosophy.

It is notorious that in asserting that oxygen was the
principle of acidity, Lavoisier was himself the
victim of the urge to generalise too soon; we must
quote further from him here, because his definition
of the term 'element' set the stage for all the
discussions of the topic in the nineteenth century:

> All that can be said, upon the
> number and nature of elements is,
> in my opinion, confined to discussions
> entirely of a metaphysical nature.
> The subject only furnishes us with
> indefinite problems, which may be

21

solved in a thousand different
ways, not one of which, in all
probability, is consistent with
nature. I shall therefore only
add upon this subject, that if,
by the term elements, we mean to
express those simple and indivisible atoms of which matter
is composed, it is extremely
probable we know nothing at all
about them; but if we apply the
term elements, or principles of
bodies, to express our idea of
the last point which analysis is
capable of reaching, we must admit
as elements, all the substances
into which we are capable, by any
means, to reduce bodies by decomposition. Not that we are
entitled to affirm, that these
substances we consider as simple
may not be compounded of two, or
even of a greater number of
principles; but, since these
principles cannot be separated,
or rather since we have not
hitherto discovered the means
of separating them, they act with
regard to us as simple substances,
and we ought never to suppose
them compounded until experiment
and observation has proved them
to be so.

There are three important points which we should
notice in this passage. The first is the denial of
the fruitfulness in chemistry of a theory of matter.
We shall find that in the nineteenth century many
chemists — perhaps the majority — took the same line
as Lavoisier, at any rate until the 1870s. A
consequence was that chemistry and physics, which had
been prematurely united by the corpuscularians, went
their separate ways to a considerable extent.
Chemists took little note of physical evidence, and
relied upon the facts of chemistry; and physicists
tended to see chemists as superior cooks or

22

apothecaries. The Fellows of the Chemical Society
of London as late as the 1860s were so unspeculative
that they refused to allow in their Journal papers
of a purely theoretical nature. As long as a
descriptive science was all that was aimed at, this
approach was not open to objection; but powerful
explanatory theories in chemistry only became a
possibility when chemists took seriously first the
chemical atom of Kekulé with its valence bonds, and
then theories of matter developed by physicists,
instead of treating them as metaphysical. Against
this, we should notice Lavoisier's determination to
make chemistry as far as possible logical and
quantitative, a genuine science deserving to be
treated with respect.

The second point is that Lavoisier provided the
famous definition of a chemical element which was to
persist in use for more than a century. We may call
the definition negative-empirical, for it is
curiously based upon what cannot at present be done:
the further analysis of bodies appearing on the list
of elements. The status of elements thus defined was
unstable; within five years Davy, using the Voltaic
cell, removed potash, soda, and the alkaline earths
from the list by decomposing them, and added chlorine
to the number by establishing that previous claims to
have decomposed it could not be substantiated. The
criterion was not always even clear; a considerable
number of new elements were announced during the
nineteenth century which no longer appear on our
lists. It was particularly hard to apply it to the
rare earth metals, all of which are extremely alike
in properties. In fact, chemists did not adhere
rigidly to Lavoisier's recommendations; all —
including even Lavoisier himself — seem to have been

open to arguments from analogy, and tended not to regard as elementary a substance which closely resembled known compounds, even though it had not itself been decomposed.[28]

The third point comes in the last sentence of the passage from Lavoisier; in which he recognised that a list of chemical elements defined in a negative-empirical manner can never be more than provisional. No means of analysis had hitherto been discovered which would resolve them into their supposed constituents; but one could certainly not draw from Lavoisier's writings the conclusion that it would be mistaken to look for more-powerful new methods of analysis. And indeed in the nineteenth century, first the Voltaic cell and then the spectroscope proved the most valuable means of probing into the structure of things, and yielded a rich harvest of new elementary bodies. On top of this, there were serious problems as to the interpretation of experimental data; in particular, of deciding whether a given experiment was an analysis or a synthesis. Thus for the believer in phlogiston, the calx was an element and the metal a compound of the calx with phlogiston. For Lavoisier, the metal was an element and the calx its oxide. Lavoisier's paradigm had to be accepted as well as his experimental data for one to agree with him about elements. It was over the status of the Aristotelian element water that controversy raged about 1800, though as we shall see earth — or earths — and air and fire, with which the chemical revolution had begun, were also involved.

NOTES

1. H.Davy, Works, VIII,pp.323-8. W.A.Smeaton,
'Schwediauer, Bentham and Beddoes: Translators of
Bergman and Scheele', Annals of Science, XXIV (1968)
273; A.L.Lavoisier, Elements (1790) p.xxi. For the
view that every science must contain a speculative
and a practical part, see I.Barrow, Mathematical
Learning (1734)pp.50ff.

2. In the early nineteenth century, W.H.Wollaston
and W.Prout were physicians; earlier C.W.Scheele
had been an apothecary; Thomas and William Henry
were manufacturers; Joseph Priestley, Stephen Hales
and R.J.Haüy were clergymen of various denominations;
Henry Cavendish a gentleman of leisure; and Thomas
Beddoes, Smithson Tennant, and Davy held academic
posts, as did Joseph Black and Thomas Thomson.
Correspondence and articles in the Chemical News in
the 1860s and 1870s casts light on the emergence of
chemistry as a profession in England — see also the
article on this by B.W.G.Holt, British Journal of
Sociology, XXXI(1970) 181-99. Professional status
came earlier in France; M.P.Crosland, The Society of
Arcueil,1967, and J.T.Merz, European Thought (1904)
chaps.1-3. On the Henrys, see the papers by W.C. &
K.R.Farrar and E.L.Scott, Ambix,XX (1973) 183-208,
XXI (1974) 179-207, 207-28, XXII (1975) 186-204;
XXIII (1976) 27-52; XXIV (1977) 1-26.

3. P.J.Marshall (ed.), The British Discovery of
Hinduism in the Eighteenth Century (Cambridge,1970)
pp.198,200,262; K.Raine and G.M.Harper (eds.),
Thomas Taylor the Platonist, Selected Writings,(1969)
p.v. Contrast D.Hartley, Observations on Man (1749)
II, p.255.

4. I.B.Cohen and H.M.Jones, Science Before Darwin
(1963) p.7; see D.M.Knight, 'The Scientist as Sage',
Studies in Romanticism, VI (1967) 65-88, and 'The
Physical Sciences and the Romantic Movement',History
of Science IX (1971) 54-75, and 'Chemistry,
phisiology and the Romantic Movement', Durham
University Journal, LXIV (1972) 139-45, and 'German
Science in the Romantic Period' in M.P.Crosland (ed.)
The Emergence of Science in Western Europe (1975).
See also B.S.Gower, 'Speculation in Physics', Studies
in Hist. & Phil. of Science, III (1973) 301-56;
I.Inkster, 'Science and the Mechanics' Institutes,
1820-50', Annals of Science, XXXII (1975) 451-74;
B.Sinclair, Philadelphia's Philosopher Mechanics,
Baltimore,1974.

5. T.Bergman, A Dissertation on Elective
Attractions (1785); reprint, intr. A.M.Duncan (1970).

6. R.E.Schofield, Mechanism and Materialism
(Princeton,N.J.,1970); A.Thackray, Atoms and Powers
(Cambridge,Mass.,1970)· B.J.T.Dobbs, The Foundations
of Newton's Alchemy, or 'The Hunting of the Greene
Lyon', (Cambridge 1975). On atomism in chemistry, see
D.M.Knight, Atoms and Elements, (1967,2nd edn.1970),
and the collections of papers Classical Scientific
Papers: Chemistry, 1st and 2nd series, (1968 and 1970),
D.M.Knight (ed.), and Physics (1964), S.Wright (ed.);
and B.Schonland, The Atomists (Oxford,1968) for a
more 'positive' view.

7. See H.W.Piper, The Active Universe (1962);
T.McFarland, Coleridge and the Pantheist Tradition
(Oxford,1969); T.L.Hankins, Jean d'Alembert: Science
and the Enlightment (Oxford,1970).

8. See Newton's letter to Bentley in I.B.Cohen (ed.),
Isaac Newton's Papers and Letters (1958) p.302; and
Tyndall's Faraday as a Discoverer in The Royal
Institution Library of Science, Physical Sciences,II,
p.79; J.Priestley, Disquisitions (1777). See the
paper by D.M.Knight in the Actes du XII[e] Congres
International d'Histoire des Sciences,1968 (1971) VI,
pp.49-53.

9.R.Boyle,The Sceptical Chymist (1661); Works, ed
T.Birch (1772); M.Boas, Robert Boyle and Seventeenth-
century chemistry (Cambridge,1958),as M.B.Hall,
Robert Boyle on Natural Philosophy (Bloomington,
Indiana,1965); R.H.Kargon, Atomism in England from
Hariot to Newton,(Oxford,1966); J.F.Fulton, A
Bibliography of ... Robert Boyle, 2nd edn.(Oxford,
1961); A.Thackray, John Dalton (Cambridge,Mass.,1972);
A.Debus, The English Paracelsians (1965); Science and
Education in the Seventeenth Century (1970); and his
forthcoming Chemical Philosophy. Aristotle's views
may be found in the de Generatione et Corruptione,
329a24-331a6; H.Cassebaum and G.B.Kauffman, 'The
analytical concept of a chemical element' Annals of
Science, XXXIII (1976) 447-56.

10. Leicester H.M. and Klickstein H.S. A Sourcebook
in Chemistry 1400-1900 (New York,1952) p.20.

11. We might notice the dialogues of Galileo, van
Helmont, Fontenelle, Berkeley, Réaumur, Hume and
Mandeville; by the nineteenth century we only find
dialogues in elementary textbooks like Mrs Marcet's
immortal Conversations on Chemistry (1806).

12. R.Boyle, Works, I, pp.467, 469, 477, 544.

13. R.Boyle, Works, I, p.548; cf W.Charleton,
Physiologia Epicuro-Gassendo-Charltoniana (1654)
bk.III, chap 1.

14. R.Boyle, Works, I, p.562; cf p.584; see Boas,
Boyle (1958) p.96; N.Grew, The Anatomy of Plants
(1682) p.224; A Thackray, "Matter in a Nut-shell",
Ambix,XV (1968) 29-53; for Newtonian atomism,
I.B.Cohen, Franklin and Newton (Philadelphia,1956);
D.S.L.Cardwell (ed.) John Dalton and the Progress of
Science (Manchester,1968); R.S.Fleming, 'Newton,
Gases, and Daltonian Chemistry', Annals of Science,
XXXI (1974) 561-74.

15. H.Davy, Works, VIII, pp.329-31; cf C.Linnaeus,
Systema Naturae (Louvain,1735) p.1 observation 15;
'Lapides crescunt. Vegetibilia crescunt et vivunt.
Animalia crescunt, vivunt & sentiunt. Hinc limites
inter haecce Regna constituta sunt.' M.Faraday, A
Course of Six Lectures on the Non-metallic Elements,
ed.J.Scoffern (1853) p.7.

16. J.E.McGuire, 'Force, Active Principles, and
Newton's Invisible Realm', Ambix, XV (1968) 154-208;
I.B.Cohen, Newton's Papers and Letters, (1958)
section 3; A.R. & M.B.Hall, Unpublished Papers of
Newton (1962) section 3; and especially A.Thackray,
Atoms and Powers (1970); P.M.Heimann, 'Newtonian
Natural Philosophy and the Scientific Revolution',
History of Science II (1973) 1-7.

17. I.Newton, Opticks (New York,1952); Davy's
inaccurate quotation is in Works,VIII, p.287;
cf J.Dalton, in Phil.Mag.,XIV (1802) 169.

18. I.Newton, Opticks, p.394; see also pp.380,382-3,
and 397; W.Thomson, Popular Lectures (1889-94) I,
pp.59ff; P.S.Laplace, System of the World (1809) II,
pp.235ff.

19. J.R.Partington, A History of Chemistry (1962) II,
pp.631,665; Thackray, in Atoms and Powers (1970)p.84
makes further suggestions as to why Newtonian ideas
did not spread among Germans.

20. W.Whewell, Philosophy of the Inductive Sciences
(1847) I,pp.383,387; With the view of Whewell — and
of Hegel — that the sciences must be kept distinct,
contrast that of Baden Powell that they are, or
should be unified; Essays (1855) pp.40-81; see paper
by D.M.Knight 'Baden Powell and the Inductive
Philosophy', Durham University Journal,LX (1967-8)
81-7; and 'The Vital Flame', Ambix, XXIII (1976)5-15;
R.Cotes' preface may be found in I.Newton,

Mathematical Principles (1962) I,pp.xx-xxxiii; Davy,
in his Syllabus of 1802, p.1, took for granted a
corpuscularian chemistry.

21. Leicester and Klickstein, Sourcebook, pp.56-7,
59, 61-2, (1691), Becher — pp.59 (1697), 61-2 (1723),
Stahl; see M.Fichman, 'French Stahlism and chemical
studies of air, 1750-1770', Ambix,XVIII (1971)94-122;
and D.Oldroyd, 'Stahl's Philosophical Principles',
Ambix, XX (1973) 36-52.

22. C.W.Scheele, 'The Discovery of Oxygen', Alembic
Club Reprints, VIII,1952, pp.5,6; cf Scheele's
Chemical Essays (1786) pp.259ff, 337ff.

23. On Hales' influence, see H.Guerlac, Lavoisier —
the crucial year, (Ithaca,N.Y.,1961) especially
pp.25-35; A.E.Clark-Kennedy, Stephen Hales,(Cambridge
1929) chaps. 6 and 12; J.Parascandola and A.Ihde,
'History of the Pneumatic Trough', Isis, LX (1969)
351-61.

24. P.J.Macquer, Elements of Chemistry, 1775,I,p.2.

25. H.Davy, Works,VII, p.115. On Priestley, see
R.E.Schofield, A Scientific Autobiography of Joseph
Priestley (Cambridge,Mass.,1966). J.Priestley, 'The
Discovery of Oxygen'. Alembic Club Reprints, VII
(1961) pp.6, 16.

26. J.Black, 'Experiments upon Magnesia Alba',
Alembic Club Reprints, I (1898) p.30; T.Cochrane,
Notes from Doctor Black's lectures on chemistry
1767-8, ed. D.McKie (Wilmslow,Cheshire,1966)
pp.xviii, 72-4; J.Robison (ed.), Black's Lectures
on Chemistry (1803) I, pp.282, 284; see also p.vii.
T.Reid, Essays on the Intellectual Powers of Man,
intr. B.A.Brodie (Cambridge,Mass.,1969) pp.46, 119ff,
145, for emphasis on 'experience' and avoidance of
'systems' by a prominent contemporary and friend of
Black.

27. See M.Boas in M.Clagett (ed.), Critical Problems
(Madison,Wis.,1959) p.511. A.L.Lavoisier, Elements of
Chemistry (1790) pp. xxiii, xxiv. One might compare
Lavoisier's remarks on his predecessors with those of
the reformer of geology, Charles Lyell; they appear
in the first volume of his Principles of Geology
(1830), and have the same slightly indulgent and
condescending tone.

28. A.L.Lavoisier, Elements of Chemistry (1790)
pp.177-8.

2

Heat, Light, and Water

Lavoisier's system of chemistry rapidly proved
successful, although isolated phlogistonists
persisted into the middle of the nineteenth century.[1]
Thus in 1849 one Stevenson, apparently an F.R.S.,
expressed surprise that no professors of chemistry
had come out in support of his book proving the
existence of phlogiston and the elementary nature of
water; since his arguments were irrefutable, he could
only suppose that vested interests must have lain
behind this pedagogical inertia. As had happened with
gothic architecture, this phlogiston survival was
almost contemporaneous with the phlogiston revival, in
William Odling's sympathetic account in 1871 of the
theory as an anticipation of the doctrine of chemical
potential energy — Odling being one of the leading
theoretical chemists in Britain.

Lavoisier's views on the chemical elements were
also generally circulated with his system; but authors
even of standard works made various reservations as to
the status of such elements, and some refused to allow
that discussion of the ultimate constituents of matter
must be metaphysical. Fourcroy,[2] echoing Lavoisier,
remarked that modern chemists were careful not to
confuse their simple bodies with the elements or

principles of the old-fashioned chemistry. Gren,[3]
on the other hand, refused in his textbook to accept
Lavoisier's guidance, writing:

> The first original substances of
> which bodies consist, and which
> are not themselves compounded of
> heterogeneous parts, are called
> elements, first principles,
> radicals. Many substances cannot
> be further decomposed by the
> chemist into constituent hetero-
> geneous parts; but this does not
> entitle us to rank them among the
> elements. Though they are as yet
> undecomposed, it does not follow
> that they are undecomposeable; as
> perhaps neither our senses nor our
> instruments will ever reach those
> substances which by their nature
> admit of no sort of decomposition.

Gren was attracted by schemes biased towards chemical
dynamics and 'Kantian' ideas and was, like many
Germans, unhappy about Lavoisier's antiphlogistic
system; but even Chaptal,[4] a supporter of Lavoisier's
chemistry, felt that ideally the term 'element'
should be effaced from chemistry because it gave rise
to confusion and was an unsuitable word to apply to
items on the unstable list of undecomposed bodies.

Lavoisier's list of elements contained not only
what we would consider undecomposed chemical substances but also the two imponderable fluids heat
('caloric') and light.[5] These two caused much
trouble, and in this chapter and the two following
we shall be concerned with them, and also with the
composition of water; this last was a problem which
Lavoisier had solved, but not to the satisfaction of
everybody. One of Gren's objections to the antiphlogistic system was that it did not allow
sufficiently for the role of light in chemistry; and

Davy, as we shall see, thought the same. But
Lavoisier was happy to provide an important place in
his system for heat, which played a role like that
of other chemical elements. In some reactions,
hydrogen was given off; in others, caloric: the
explanations were similar. We might feel with
William Nicholson — in the preface of his translation
of Fourcroy — that too many chemists were too ready
to consider heat and light as capable of entering
into combination like other bodies, and of 'being
disengaged by the action of the affinities".

In the later chapters we shall look at ideas of
the nature of heat and light as they relate to
materialism and as they indicate Newtonian survivals
in chemistry; here we need therefore only examine
them in their chemical applications, with particular
reference to ideas on the composition of water.
Because combustion was the only exothermic reaction
with which chemists of this period were familiar, the
evolution of heat in chemical reactions was supposed
by Lavoisier and his successors to be connected with
oxygen. Thus the caloric theory of heat and the
oxygen theories of burning and of acids were linked,
and until after Davy's work on chlorine, chemistry
was a combustion-dominated science. For Lavoisier,
oxygen gas was a compound of oxygen — an unknown
basis — and caloric, by which he meant, for example,
that since sulphur is a combustible body it 'possesses
the power of decomposing oxygen gas, by attracting
the oxygen from the caloric with which it was
combined'.[6] Calorimetric experiments might enable us
to quantify these attractions.

Lavoisier allowed that light had a chemical role;
and Fourcroy, who had supported a kinetic theory of

31

heat but came round to Lavoisier's view, added the
fresh speculation that since reactions with oxymuriatic acid (our chlorine) produced less light
than those with vital air (our oxygen), the latter
gas must contain more light than the former.[7]
Humphry Davy's first published paper was on the same
topic; he supported the kinetic theory of heat by
experiments, and argued for the presence of combined
light in various chemical substances. Davy
originated from Penzance in Cornwall, then a kind of
Wild West with wreckers and smugglers as well as
mining and steam-engines; he was born in 1778, a few
days after Sir Joseph Banks, whose immediate success-
or Davy was to be, had been elected President of the
Royal Society. He wrote later that he was glad that
he had not been pushed too hard at school; he left
early and was apprenticed to an apothecary-surgeon.
He embarked upon an extensive course of self-
education, including languages, theology, and the
sciences; but when he came to chemistry, using
Lavoisier and Nicholson as his guides, he dropped the
rest of the programme and began research. His famous
ice-rubbing experiment was done at this stage; the
results are apparently impossible, but his intuition
was right and he seems to have adhered to the kinetic
theory of heat through his life.

Ideas on light not unlike Davy's had been
published anonymously in a series of essays by
gemtlemen at Exeter in 1796; Exeter being at this
time, so Disraeli tells us, a literary metropolis.[8]
The author of the essay held a corpuscular theory of
light, believing that the phenomena of
phosphorescence proved light to be composed of
mutually-repulsive particles, which might become

latent and quiescent. He listed various chemical
reactions of light, including the blackening of
silver salts; and knowing the importance of light
for the growth of plants, he thought that it might
reach the roots in the form of electricity, 'a
peculiar modification of light'. While ideas like
Davy's — derived ultimately from Newton's corpuscular
theory of light — were thus circulating in his part
of the country, it seems probable that as Hartley
suggests Davy's experiments were the result of reading a purple passage in Lavoisier.

There is an irony about this, for Davy believed[9]
that his ice rubbing experiment demonstrated that
while Lavoisier's theory 'will ever be admired by
chemical philosophers', it was possessed of two
defects: 'the assumption of the imaginary fluid
caloric, and the total neglect of light'. Davy
established the difference between heat and light by
flashing a gunlock in vacuo, when heat was produced
but no light was visible. Heat could therefore be
motion, while light was matter. On Davy's view,
oxygen gas was not a compound of caloric for there
could be no such things; it was a compound of light
and oxygen, to be called 'phosoxygen' — a term
expressing 'a chemical combination of the simple
substance light with the simple substance oxygen'.
The particles of light were so small as to be
scarcely affected by gravitation, and moved with
almost inconceivable swiftness. Light was an
essential constituent of living bodies; to understand
its combinations was of infinite importance, for upon
them perception, thought, and happiness depended.
Even electricity, he suggested in a passage omitted
from the published paper, might be the result of the

condensation of light.

These papers were published, when Davy was twenty, by Thomas Beddoes in a collection with the title Contributions to Physical and Medical Knowledge. Beddoes had been Reader in Chemistry at Oxford, but had resigned after supporting the French Revolution and criticising the Bodleian Library. He was the translator of Bergman and Scheele, and the author of medical works and of a best-selling moral tale. At Oxford his lectures had attracted audiences larger than any since the thirteenth century; he knew members of the Lunar Society of Birmingham;[10] and he had literary connections, for his wife was an Edgeworth, and their son was a poet. Following the 'Brunonian' doctrine of irritability, he prescribed opium to Coleridge; and also began experiments on the administration of factitious airs — synthetic gases — in the cure of diseases. He appointed Davy as his assistant in these researches, at Bristol, writing to Erasmus Darwin that he seemed 'the most extraordinary person that I have seen, for compass, originality, and quickness of thought'.[11]

In a paper which appeared with Davy's in the Contributions, Beddoes discussed the status of the chemical elements:

> Concerning the composition of the metallic substances, not much can be said. Whether to create a diversified system of bodies out of one, or out of a few or many elements, imply most wisdom or power, is a question which different persons would decide according to their various taste in world-making.[12]

Beddoes thought that the dispute might go on for as many centuries as had already elapsed since it was

begun by Thales of Miletus; but there were some
indications that the metals were complex, and by
'accurate close experiments on organised bodies'
this point might be proved. The existence of iron
and manganese in plants and animals suggested that
these metals were compounded by the organic powers;
analogy would then suggest that the other metals
consisted of the same principles, differently
modified.

In support of this belief Beddoes quoted John
Abernethy's Essays.[13] Abernethy was a fashionable
doctor, and F.R.S. He belongs to the tradition,
which began with van Helmont, according to whom the
four ancient elements were transmuted by organisms,
and perhaps could be transmuted in the laboratory.
This tradition, with its basis in both artificial and
natural observations, is well described by Bishop
Watson, whose book merited the approval of Gibbon and
was to arouse the interest of Odling. In artificial
experiments described by Watson, water was converted
into earth by prolonged heating; Lavoisier showed,
though not to Watson's complete satisfaction, that
this 'earth' came from the still and not from the
water. The natural experiments were those like that
of Cusa and van Helmont, in which a tree was grown
from water alone; Abernethy grew various vegetables
in this way, and determined to see whether animals
could grow in pure water. He tried with tadpoles;
but his weekly filtration was not efficient enough to
keep away scum, and he failed to rule out cannibalism.
On another tack, finding that water-grown plants
yielded more ash than an equivalent number of seeds,
he concluded that minerals had been synthesised.
Similarly, analyses of new born chicks showed that

35

their ashes were heavier than those of eggs. An experimental rabbit fed on water-grown vegetables grew, though nourished 'only by a modification of air and water, from which his organs are capable of forming the mineral alkali, phosphorus, and every substance which formally exists in the animal body.' Abernethy had begun these researches because he:

> had imbibed the opinions of the
> great philosophers of this island,
> who, from reflection and reasoning,
> were induced to believe, that the
> ultimate particles of matter were
> the same, and that various
> substances, with which this world
> presents us, were only differences
> in the arrangement and motion of
> similar particles.

Beddoes realised that such ideas were still conjectural; but he hoped that progress towards decomposing the metals would on the one hand yield new arts, and on the other open the way to 'chemical physiology' and hence to a genuine science of medicine.[14] Unlike Davy, Beddoes did not succeed in putting his ideas to some kind of experimental test; he knew his weaknesses, and there is a moving letter from him to Davy, written just before his death, deploring his lack of concrete achievement.

Davy's paper on light and heat seems to have converted only Beddoes; and he later withdrew his term 'phosoxygen', and begged 'to be considered a sceptic with regard to my own particular theory of the combinations of light, and theories of light in general'.[15] This retraction should not, as we shall see, be taken as an abandonment of theorising; his later papers contain a mixture of sober reporting and speculation. Indeed, he can be accused of habitually proposing in general terms a number of

36

incompatible explanations of phenomena, so that
whatever turned out he could claim to have been
right. Even in 1799 the Philosophical Magazine
could greet his appointment to Beddoes' Pneumatic
Institution with the words; 'it would hardly be
possible to find one better qualified for the
situation'; but Davy's reputation was made in 1800
when he demolished the theory of the American
physician Samuel Mitchill that nitrous oxide was the
principle of contagion, which must bring immediate
death to any who respire it. Davy discovered the
anaesthetic properties of the gas, and breathing it
became a craze; his Researches on the oxides of
nitrogen became a standard work, and helped to make
the neutral term 'nitrogen' rather than 'azote'
(implying poison) the English term for the element.
With this work just completed, Davy was in an
excellent position to take up the study of Voltaic
electricity when Volta's discovery was announced in
1800. In Britain at this time there were two
valuable general journals, The Philosophical Magazine
and Nicholson's Journal, as well as the august
Philosophical Transactions of the Royal Society: the
two private journals offered rapid publication, and
included translated articles, and reviews.

It was not surprising that the first experiments
made in England with Volta's battery[16] should have
been on its application to the decomposition of water.
At the same time as Nicholson and Carlisle did their
classic experiment, a translation was appearing in
Nicholson's Journal of the papers of Girtanner and
others on the composition of nitrogen and of muriatic
acid. Christoph Girtanner[17] was a physician who had
written a daunting practical and bibliographical work
on venereal disease, arguing for the American origin

of syphilis. He became one of the first German
converts to Lavoisier's theory, and wrote both a
standard textbook — used by Berzelius — and a German
version of the new nomenclature.

In the <u>Philosophical Magazine</u>[18] we find an
account of this controversy about nitrogen, which
like many chemical arguments of this period had
nationalistic overtones. Van Mons wrote from
Holland that Girtanner had analysed nitrogen and
shown it to be 93% hydrogen and 7% oxygen; figures
later corrected by Girtanner to 79% and 21%. Water,
air, ammonia, and nitrogen were therefore all compounds of oxygen and hydrogen, and in analysing air
we do not separate but synthesise the nitrogen. Van
Mons had failed to repeat Girtanner's analyses; and
he was unhappy about them as he was about the view
that muriatic acid (our hydrogen chloride) was also
an oxide of hydrogen. Girtanner had passed steam
through a hot tube, and obtained nitrogen; German
chemists concluded that Lavoisier was thus proved
wrong, and that water was the ponderable base of all
gases. Their explanation of his experiment was that
water + heat = nitrogen: a view compatible both with
the phlogiston theory, still being defended by
Priestley, and with the 'dynamical' views of Ritter
and Winterl. The Dutch chemists, anxious to preserve
Lavoisier's theory, denied that steam could be
converted into nitrogen. Girtanner sought the best
of both worlds; he was sorry to see 'party spirit'
and 'asperity' intruding upon the discussion, and
hoped that if all adhered to facts rather than
systems then some truth would be found on both sides.
He found that steam was converted into nitrogen when
water was boiled in unglazed vessels, or passed

through unglazed pipes, or boiled with quartz, lime, or silex. Under other conditions it was not; so nitrogen cannot simply be water + heat, but is rather water deprived of some of its oxygen, which has been absorbed by the 'earth'.

Girtanner found that tin and water yielded an oxide and nitrogen; and recalled that Scheele when he breathed hydrogen in and out of a bladder a few times found nitrogen there, proving that hydrogen with some oxygen did produce nitrogen. Nitrogen was therefore correctly named 'azote'; it was a poison because of its affinity for oxygen. Eudiometry was founded upon erroneous principles; and the synthesis of saltpetre must be further investigated. He concluded that no doubt in the nineteenth century transmutation would be understood and practiced. Although Van Mons was sceptical about the synthesis of nitrogen, he produced various arguments[19] from analogies for the suggestion that potash was a compound, containing nitrogen.

It is by comparison with such experiments as those of Girtanner that we can judge how competent were Davy and Berzelius a few years later. Girtanner's procedures were not very good; but no experiment could establish whether water is beyond doubt an element or a compound, because this is a matter of interpretation. We should remember that these papers, and those of Mitchill,[20] appeared in respectable journals; chemistry was in a revolutionary phase, and its very foundations might be shaken as they had been by Lavoisier. No axioms were so deeply entrenched as to be exempt from attack; and, like Davy, one might hope to make important and original contributions within months of taking up the science. As William Whewell later wrote of electrical science:

> At such a period, a large and
> popular circle of spectators
> and amateurs feel themselves
> nearly upon a level, in the
> value of their trials and
> speculations, with more profound
> thinkers; at a later period, when
> the subject is become a science,
> that is, a study in which all
> must be left far behind who do
> not come to it with disciplined,
> informed, and logical minds, the
> cultivators are far more few,
> and the shout of applause less
> tumultuous and less loud. We may
> add, too, that the experiments,
> which are the most striking to
> the senses, lose much of their
> impressiveness with their novelty.[21]

Chemistry was in 1800 still in this first stage;
lecturers attracted large audiences and amateurs
tried chemical experiments. Davy recommended a
liberal education, and qualities of patience, neatness,
industry and accuracy, for the would-be chemist; but
warned that 'the business of the laboratory is often
a service of danger', and recommended beginning where
'the path is marked and distinct'.[22]

Into this hotbed of speculation and activity
came in 1800 Volta's announcement of the electric
battery. Its impact was immediate, and there was an
almost frenzied eagerness among chemists to work with
this new device, which was so easy to make; Nicholson
and Carlisle used half-crown coins for their silver
discs, and salt water for their medium. Ten years
later, when order had been brought into this region
of science, Davy remarked that:

> The voltaic battery was an alarm
> bell to experimenters in every
> part of Europe; and it served no
> less for demonstrating new properties
> in electricity, and for establishing

the laws of this science, than
as an instrument of discovery
in other branches of knowledge,
exhibiting relations between
subjects before apparently without connection, and serving as a
bond of unity between chemical and
physical philosophy.[23]

In chemistry its chief value at first was to be as an
analytical tool of hitherto unapproached power; but
not until after it had been established to general
satisfaction that the battery was an agent of analysis
and not a generator of new substances. This
necessary step was taken in the investigations of the
effect of an electric current flowing through water.

Nicholson and Carlisle showed that under these
circumstances hydrogen came off at the negative pole,
and oxygen at the positive; this did not of course
prove the decomposition of water, for oxygen and
hydrogen might well be electrically-charged forms of
water. But whatever explanation was favoured, the
first essential was to establish that the Voltaic
electricity was analogous to the common Franklinic
electricity produced by rubbing. The identify of
Voltaic, Franklinic, and animal electricity — from
electric fish — was only put beyond doubt by Faraday
in 1833; but in 1801 William Hyde Wollaston had shown
parallels between Voltaic and Franklinic electricity,
and Cavendish had earlier made a model electric fish
with a Leyden jar to show that animal and common
electricity were one.[24] Wollaston had first to
investigate the origin of the electricity. Volta
suggested that mere contact of dissimilar metals was
enough to generate it; but in Nicholson's Journal, in
the same number as Nicholson and Carlisle's experiment,
both Davy and one Haldane had suggested that the
current was maintained by a chemical reaction. Davy

noted that the pile only worked:

> when the conducting substance
> between the plates is capable
> of oxidating the zinc; and that
> in proportion as a greater
> quantity of oxygen enters its
> combination with the zinc in a
> given time, so in proportion is
> the power of the pile to
> decompose water and to give the
> shock greater ... the oxidation
> of the zinc in the pile, and the
> chemical changes connected with
> it are <u>somehow</u> the cause of the
> electrical effects it produces.

Wollaston accepted that the quantity of electricity was proportional to the amount of chemical change; but remarked that it was not clear whether the chemical change produced the electricity or whether the latter, produced by contact, occasioned the chemical reaction. To us, the most interesting part of his paper was his decomposition of water by frictional electricity. This does not seem to have been quite new; ten years earlier, Erasmus Darwin had written that 'electricity decomposes water into these two airs, termed oxygene and hydrogene', and Michael Donovan names three Continental authors who had anticipated Wollaston.[25] The problem is to achieve a high current density; Wollaston did this by using very small poles, made by sealing fine gold wires into glass and filing the glass away until a gold point was visible through a lens. He estimated the diameter of the points as between 1/1500 and 1/700 of an inch; and found that when the poles were dipped into water and connected to the electrical machine, a mixture of oxygen and hydrogen was given off at each pole. From this similarity to Voltaic electricity — a similarity Faraday refused to allow, for with the latter the

gases come off separately — Wollaston concluded that
the two electricities must arise from the same
source, and even suggested that there might be oxidation in the frictional machine. In 1810, Davy
described this paper as 'the first distinct
demonstration of the identity of [the battery's]
effects with those of the common machine'.

Most of the early work was less successful than
Wollaston's. On the Continent[26] Johann Ritter made
the distinction between the intensity and the
quantity of electricity, to explain why the effects
of four pairs of plates differ from those of one
pair or four times the area even though the amount of
chemical action must be the same. Ritter also tried
to charge a Voltaic battery with an electrical
machine; and passed electricity through himself for
an hour and got an attack of diarrhoea. Ritter's
creative life was short; and the most significant
work done on the Continent was that done by Berzelius
in collaboration with Hisinger. They reached
conclusions similar to those later announced by Davy,
and must share some of his honours; but their papers
were not generally known until after Davy's Bakerian
Lecture of 1806 was published, and his conclusions
were reached quite independently.

Elsewhere,[27] we find the French chemist Desormes,
who in general supported the oxidation theory,
discovering acid and alkali produced around the poles
during electrolysis and thus weakening the view that
electricity merely decomposes bodies. While few
people seem to have supposed electricity to be an
element like heat or light, M. Robertson, a supporter
of Volta's contact theory, announced in France his
discovery of a series of salts of electricity, which
he called galvanates, their parents being galvanic

acid. The eminent investigators Fourcroy,
Vauquelin, and Thenard showed that a pile with more
plates decomposed water faster, whereas with larger
plates a better heating and sparking effect was
attained; whilst Biot and Cuvier established that
oxygen was necessary for the continuing action of
the pile. Biot thought that the electricity was
produced by contact, and that oxidation merely
provided 'a more intimate means of communication'
between metals.

Work in England was also inconclusive for the
most part, except ultimately for that of Davy. In
1800, he was still at Bristol, and his first
experiments were performed on Beddoes' battery of
110 double plates; but in 1801 he was appointed to a
lectureship at the Royal Institution in London, and
soon became Professor there. In the opening years of
the century we find in Nicholson's Journal and in the
Philosophical Magazine — the two journals at which the
historian of this period will look first — numerous
papers of varying importance, some anonymous and
others signed by persons of varying degrees of
expertise. In 1802 the elderly Joseph Priestley[28]
wrote from America defending phlogiston, and giving a
phlogistic account of the emission of hydrogen and
oxygen on electrolysis. He considered 'the modern
hypothesis of the decomposition of water', though
almost universally received, to be 'wholly chimerical,
and unable to stand its ground much longer'. He
proposed that there were two electric fluids, the
positive containing the principle of oxygen, and the
negative that of phlogiston; united to water, these
constituted dephlogisticated and inflammable air, our
oxygen and hydrogen. Electricity and phlogiston were
similar, and one need suppose no electric current;

the calcination of the zinc provides phlogiston, and
when that ceases so does the action of the pile.
Both ends of the pile might be made, in his view, of
the same metal; but if oxygen and hydrogen were
produced by electrolysis in vacuo this might prove
correct the 'new theory' of Lavoisier.

Priestley's view was shared at various times by
Davy and Ritter; but the little-known Sylvester[29]
wrote in 1804 that it was more gratuitous to assume
two electric fluids than to concede that water was
being decomposed into its elements. The two fluids
would continually meet and neutralise each other in
the water, or in the wires between cells, and
generate caloric. Similar worries about motions of
opposite particles were voiced by 'a correspondent'
in 1801; since the oxygen and hydrogen appeared
separately, 'If a particle of water is composed of a
particle of oxygen and a particle of hydrogen, then
what rapid currents must there be of those two
substances!'. He believed therefore with Priestley
that water was the basis of these gases; Lavoisier's
'basis of oxygen gas' was a nonentity.

For this correspondent, positive and negative
electricity were distinct fluids; 'Governor' Pownall,
late of Massachusetts, tried to penetrate further.[30]

Assuming Newton's aether (insubstantial matter
that might explain gravity) 'as a datum, as a known
fact', he employed it as a basis for explaining
electricity, light, gravity, chemical affinity, and
magnetism; one sees a vista of the whole of
nineteenth-century physics, but Pownall was no
Maxwell and his aether does not anticipate the electro-
magnetic field. In the same year the attention of
'Mr. Davy and other philosophers' was drawn by another
anonymous correspondent to the problem of intensity

45

and quantity of electricity from the cell. Nicholson
attacked the phlogistic theory (put forward on this
occasion by Gibbes, a physician from Bath) on the
grounds of polarisation, asking why oxygen from the
air should be necessary to keep the cell going if
oxidation were simply the combination of metal and
water, with a separation of electricity. Meanwhile,
Wilson explored the contact theory, shaking chips of
one metal through a sieve of another and seeing what
charge they acquired. In 1806 one H.B.K. reported
the production of nitrous acid during the electrolysis
of potassium carbonate; and Thicknesse, from a famous
landed family, added to the available hypotheses the
compromise view that water is 'composed of oxigen,
hydrogen, and the electric fluid'.

In 1802 there appeared an historical sketch of
galvanism by John Bostock,[31] who followed it by a
theory of his own. The experiment of Nicholson and
Carlisle he considered 'as the most important since
the discovery of oxigen by Dr. Priestley'. This was
tactful of him in that he was writing for Nicholson's
Journal, but we can concede his point while allowing
for some patriotic exaggeration. He mentioned the
work of Haldane, who had shown that the pile would
not work in an atmosphere of nitrogen and went best
in oxygen, providing evidence for a chemical theory
of its action. Davy had, in Bostock's view, made
'the first step towards the true theory of the action
of Volta's pile'; poor Volta himself was evidently
'un-acquainted with many of the most interesting
circumstances which had been observed in this country,
respecting the operations of the apparatus originally
designed by him', and which confuted his theory.

Davy observed in 1810[32] that he scarcely knew
how to describe with propriety the distinguished

46

labours of his countrymen, which could, in 'strict
philosophical justice', be compared with all that the
Continental scientists had achieved. Some of the
most remarkable chemical phenomena connected with the
battery had been discovered in Britain soon after its
invention. It would be indelicate, he said, to single
out individuals for praise:

> Their merits are universally
> acknowledged; concerning the
> development of new philosophical
> facts there can be but one
> judgment, for facts are
> independent of fashion, talk,
> or caprice, subject to no code
> of criticism, the creations of
> truth, and durable as that nature
> to which they belong.

In fact, as Davy and his audience were well aware,
most of the important discoveries made in England
had been made by him; the main exceptions being the
papers of Nicholson and Carlisle, and of Wollaston.

Davy published in 1800 a series of papers
verifying that oxygen and hydrogen are given off from
water;[33] he obtained too little oxygen, as others did,
but concluded that this happened because the
solubility of oxygen in water was greater than that
of hydrogen. The purer the water, the better were
the proportions; but with really pure water in the
pile he found that it would not work. He accordingly
adopted the chemical theory, and tested it by using
acids instead of water between the plates. Concentrat-
ed nitric acid was the most effective medium; so
effective indeed that he had to throw the pile into
water to prevent its destruction, and got a shock in
so doing. When acids were used, he found that the
pile would operate in vacuo, and that oxidising acids
prevented polarization. As a further blow to the

47

contact theory, he found that two metallic poles are not necessary; a battery with one pole of charcoal and another of zinc was quite effective. He found further that electrolysis proceeded happily when the poles were in separate cups, joined by bridges of moist amianthus (a woven asbestos fibre);[34] and much of his later work was done in apparatus of this kind. The circuit could be completed by people holding hands; Bostock reported that in this case women got a worse shock than men. In a lecture at the Royal Institution on 1 September 1801 Davy divided conductors into two classes; the 'perfect' conductors were metals and charcoal, and the 'imperfect' were 'oxydated fluids' (acids) or substances containing them. It was a sufficient condition for galvanic action that we have 'at least two bodies of the same class and one of a different class'; batteries could thus be made with two fluids and one metal, attacked by one of the fluids, and Davy made one. This experiment, as Nicholson remarked, quite overthrew Volta's theory. Other papers by Davy at this period included descriptions of the electric arc, and discussions of the nature of electricity and its relationship to galvanism. That galvanism was electricity, he said in 1802, was a view 'adopted by the greatest number of those philosophers who have studied the subject with accuracy. Galvanism is now believed to be electricity; and the chief difference that prevails in theory is with respect to the manner in which this electricity is excited.'

We have now followed Davy up to the time of Bostock's article; and also to the date at which Davy seems to have more or less dropped electrical experiments for about four years. In 1803 he was

elected a Fellow of the Royal Society; he seems at
this time to have devoted much of his time to
applied chemistry. Bostock's theory,[35] presented at
the end of his historical account, was based upon
three postulates: that electric fluid was given out
during oxidation, that it has a strong attraction
for oxygen, and that it passes invisibly through
water, combined with hydrogen. He hoped thus to
conserve the strongest parts of both the oxidation
and contact theories; the metals must be such as to
produce a current on contact, and the fluid such as to
oxidise one of them, if a current is to flow. Bostock
was critical of Wollaston for pushing the analogy
between the voltaic cell and the electrical machine
further than it would go; in his opinion, the machine
merely collected electricity, while the cell generated it. He believed that given his three postulates
he could explain all the facts set out in the
historical section. We should notice, and it is a
point to which we shall return later, that for this
author electricity was a thing, a kind of fluid.

In the next year, Hisinger and Berzelius showed,
though little note was taken of their paper, that
pure water gave only oxygen and hydrogen on electrolysis; and Grotthus proposed a chain mechanism for
electrolysis.[36] At last some certainty, and a
fruitful hypothesis to account for the phenomena, had
been achieved. And in 1806 Davy took up the subject
again, reporting his progress in November in the
Bakerian Lecture to the Royal Society. Donovan in
1816, reflecting on this lecture, wrote:

> The facts and generalisations
> stated by Hisinger, Berzelius,
> and Grotthus, were now extended
> and illustrated by Mr Davy. It
> were a neglect of the labours

49

of this philosopher not to
declare, that scarcely any
department of the new branch
of knowledge has come under
his examination, without being
enlarged by new discoveries,
illustrated by the diffusion
of new light upon what was known,
or impressed with some mark of
his characteristic ingenuity.

The first part of this Bakerian Lecture was taken up
with establishing exactly what happens when water is
electrolysed.

Nicholson and Carlisle had reported that oxygen
and hydrogen were given off, and from his earliest
experiments Davy seems to have believed that from
pure water these gases would be the only products.
But other workers had noticed both that the ratios of
these gases were not exactly those in which they
combined to form water, and that acid and alkali
appeared around the poles. Davy[37] began his
experiments using glass vessels, but he soon switched
over to agate ones for greater accuracy. Chemical
apparatus at this date was not infrequently made of
greenish-brown bottle glass of variable composition;
even the clear glass vessels used when greater
accuracy was a desideratum were far from inert
chemically. Davy duly found that less acid and
alkali were generated when the electrolysis was
carried out in agate apparatus; and that whatever
material was used, the quantity of acid and alkali
was less on each repetition of the experiment. It
thus seemed likely that the acid and alkali were
coming from the apparatus and not from the water,
like the 'earth' in the transmutation experiments
which Lavoisier had refuted.

The acid and alkali turned out on analysis to be

nitrous acid and a 'fixed alkali'; that is, not
ammonia. To eliminate dissolved matter from the
vessels, he next had them made of gold, but got the
same results as with the agate. Suspecting his
distilled water, he redistilled it slowly from a
silver still and was indeed left with a residue of
lead nitrate and sodium carbonate. Using this redistilled water, he still got acid and alkali on
electrolysis, but the latter was now ammonia. In
his laboratory notebook[38] one of these experiments is
recorded:

> The two agate tubes every
> precaution being taken after
> 24 hours acid & alkaline but
> the acid did not affect Nitrat
> of Silver nor Acetate of Barytes
> but acted on silver but (?) here
> is every reason to believe nitrous
> acid — by much (?) likewise The
> alkali did not appear more vivid
> during evaporation but rather less
> vivid — & left scarcely a sensible
> residum which had no effect upon
> turmeric or even litmus!!!

He added pieces of glass to the liquid and found that
acid and alkali appeared more rapidly, proving that
the glass was dissolving.

Davy thought that the acid and alkali produced
from pure water in the absence of glass must have come
from nitrogen in the air combining with the oxygen
and hydrogen respectively. On performing the electrolysis in vacuo, he found that the water still showed
slight acid and alkaline properties, but that the
alkali disappeared overnight, indicating that it was
ammonia. This could only have come from nitrogen
dissolved in water. Priestley had shown, rather
curiously but to his satisfaction, that hydrogen
expels nitrogen from water whereas oxygen does not;

and accordingly Davy found that the concentration of
acid was proportional to the time during which the
water had been electrolysed, while that of the
ammonia reached a maximum — presumably when all the
nitrogen had been expelled around the negative pole.
On 11 November 1806, therefore, Davy evacuated his
apparatus, filled it with hydrogen, evacuated it
again, filled it with hydrogen, and then performed
the electrolysis. Even after the current had been
flowing for some hours, no acid or alkali could be
detected around the poles, and Davy could at last
conclude[39] that: 'It seems evident that water,
chemically pure, is decomposed by electricity into
gaseous matter alone, into oxygene and hydrogene'.

Thomas Thomson remarked[40] of this paper: 'I
consider it not merely as the best of all his own
productions, but as the finest and completest
specimen of inductive reasoning which appeared during
the age in which he lived'. And indeed one cannot
withhold one's admiration for so clear a piece of
work, which established one fixed point in a still-
unmapped region. Although Davy said that water was
'decomposed' it was still not necessary to believe
this; indeed he himself later toyed with the view
that water was the 'basis' of oxygen and hydrogen.
But it ceased to be reasonable to believe that
electricity could produce results qualitatively
different from those of other chemical processes; it
was an agent for analysis, but not a generator of new
substances or a philosophers' stone. It was still
possible to think of electricity as a thing, a kind
of element capable of combining with other bodies,
rather than as a power; but gradually fewer people
seem to have done so.

The rest of Davy's lecture described experiments on the transport of the components during electrolysis, which were less conclusive than those on the decomposition of water; and a statement of his view of the relationship between chemical affinity and electricity; namely that they are different manifestations of the same power, the one acting upon particles and the other upon masses. Chemical bonding was electrical in nature. Polarity thus entered chemistry, although Davy was careful to avoid the term; and the logical consequence of his theory was that the voltaic cell might be used to analyse substances other than water. In the following year, Davy did so apply it, and isolated potassium from potash.

Priestley had seen the promised land,[41] writing:

Hitherto philosophy has been chiefly
conversant about the more sensible
properties of bodies; electricity,
together with chymistry, and the
doctrine of light and colours, seems
to be giving us an inlet into their
internal structure, on which all
their sensible properties depend.
By pursuing this new light, therefore,
the bounds of natural science may be
extended, beyond what we can now form
an idea of. New worlds may open to
our view, and the glory of the great
Sir Isaac Newton himself, and all
his contemporaries, be eclipsed, by a
new set of philosophers is quite a
new field of speculation.

With the work of Davy, who was duly compared with Newton, electricity and chemistry were indeed united in his bold attempts to peer into the innermost recesses of matter. Chemical theory took a new turn, and an understanding of chemical affinities seemed once again possible. The dream of chemistry as a

53

dynamical science rather than as manipulation and erudition was given fresh life. Thus Berzelius described Davy's first Bakerian Lecture as 'among the finest memoirs with which chemical theory has been enriched', and the Edinburgh Review described his second, announcing the discovery of potassium, as 'the most valuable [paper] in the Philosophical Transactions since the time when Sir Isaac Newton inserted, in that celebrated collection, the first account of his optical discoveries'. In the next chapter, we shall look at some of the implications of the view that chemical affinity and electricity were manifestations of one power; and particularly at the connections, prominent in Priestley's work, between chemical theory and materialism.

NOTES

1. W.F.Stevenson, The Composition of Hydrogen, 2nd edn.(1849); cf J.Priestley, The Doctrine of Phlogiston established, (Northumberland,Penna.,1801); W.Odling, 'On the Revived Theory of Phlogiston', Royal Institution Library of Science, II,282-91.

2. A.F.Fourcroy, A General System of Chemical Knowledge,(1804) I,p.154; Fourcroy's view was shared by his translator, William Nicholson, in his Dictionary of Chemistry (1795), 'Element'. On Fourcroy, see W.A.Smeaton, Fourcroy, chemist and revolutionary (1962). On standard textbooks and journals, see D.M.Knight, Natural Science Books in English,1600-1900 (1972).

3. F.C.Gren, Principles of Modern Chemistry (1800) p.15. On 'Kantian' and Romantic science, see the paper by B.S.Gower in Studies in the History and Philosophy of Science, III (1973) 301-56; and those by H.A.M.Snelders, 'Romanticism and Naturphilosophie and the Inorganic Natural Sciences', Studies in Romanticism, IX (1970) 193-215, and 'The Influence of the Dualistic System of Jakob Joseph Winterl on the German Romantic Era', Isis, 61 (1970) 231-40. See also L.P.Williams, The Origins of Field Theory (New York,1966); F.Szabadvary, History of Analytical

54

Chemistry,tr. G.Svehla (Oxford,1966) pp.48-50.

4. J.A.Chaptal, Elements of Chemistry (1795) I,p.62.

5. A.L.Lavoisier, Elements of Chemistry (1790)
pp.175-6. See the paper read by A.M.Duncan at the
XIIe Congress on the History of Science, Moscow 1971,
which will appear in its Actes, on whether heat was
a chemical substance in eighteenth-century chemistry.
See also R.J.Morris jun., 'Lavoisier on Fire and Air'
Isis,LX,(1969) 374-80, and the comments by H.Guerlac
which follow it; C.Perrin, 'Prelude to Lavoisier's
theory of calcination', Ambix,XVI, (1969) 140-51;
S.G.Brush, 'The Wave Theory of Heat', British Journal
for the History of Science, V,(1970) 145-67; R.Fox,
The Caloric Theory of Gases: from Lavoisier to
Regnault (Oxford,1971); R.J.Morris, 'Lavoisier and
the caloric theory', British Journal for the History
of Science, VI(1972) 1-38.

6. A.L.Lavoisier, Elements of Chemistry (1790)p.62.

7. W.A.Smeaton, Fourcroy (1962) p.104; H.Hartley,
Humphry Davy (1966) pp.15ff; Davy's paper appeared in
T.Beddoes, Contributions (Bristol,1799) and may be
found in his Works, II, pp.5-86. See the biographies
of Davy by J.A.Paris (1831) and J.Davy (1836); for
background, see R.M.Barton (ed.), Life in Cornwall in
the early 19th century, Truro, 1970; J.Z.Fullmer,
Humphry Davy's Published Works (Cambridge,Mass.,1969).

8. Essays by a Society of Gentlemen at Exeter (1796)
p.491; B.Disraeli in I.D'Israeli, Curiosities of
Literature (reprint,New York,1964) p.5; on contemporary interest in phosphorescence, see J.L.Lowes, The
Road to Xanadu, 15th printing (n.d.,Boston,Mass.)
book 1,chap.3.

9. H.Davy, Works,II,p.41; see also p.7, and pp.23
and 29.

10. R.E.Schofield, The Lunar Society of Birmingham,
(Oxford,1963).

11. J.E.Stock, Memoirs of the Life of Thomas Beddoes
(Bristol,1811) p.177 — see also p.162. This biography
includes a bibliography. See also T.H.Levere 'Dr
Thomas Beddoes', Notes and Records of the Royal
Society, XXXII (1977) 41-9.

12. T.Beddoes, Contributions (1799) p.222.

13. J.Abernethy, Surgical and Physiological Essays
(1793) pp.77, 92, 104; R.Watson, Chemical Essays,
3rd edn.(1784-7) IV, pp.257-308; see R.Cochrane,
Black's Lectures (1966) p.19 for the view that water
is converted into earth on repeated distillation. On

the element water, see the paper by Robert Southwell
of 1675 in T.Birch, History of the Royal Society,III
(1757) pp.196-216; John Woodward the geologist was
sceptical about the nourishment of plants with water
alone; Phil.Trans.,21 (1699) 193-227; that the
problem was still of interest at the beginning of
the nineteenth century is indicated by the full
account of these experiments which appeared in the
Abridgement of the Phil.Trans. by C.Hutton et al.,
(19 vols, 1809) IV, pp.382-98. A.L.Lavoisier showed,
though not to Bishop Watson's complete satisfaction,
that the 'earth' found in the distillations came
from the still; Oeuvres (Paris,1862-93) II, pp.1-28;
on a related contemporary problem, see W.A.Smeaton,
'Is water converted into air? Guyton de Morveau acts
as arbiter between Priestley and Kirwan', Ambix, XV
(1968) 73-83.

14. T.Beddoes, Contributions (1799) p.230; Views and
experiments like those of Abernethy were published
by a Mr New in Nicholson's Journal, XXIII (1809) 71.

15. H.Davy, Nicholson's Journal,4° series III (1799-
1800) 517; Phil.Mag., 5 (1799) 301. Mitchill's paper
was published as an appendix to T.Beddoes & J.Watt,
Considerations on the Medicinal Uses of the Factitious
Airs,(Bristol,1795); except in his Agricultural
Chemistry (1813), Davy favoured 'nitrogen' over
'azote' as early as his first paper, Works, II, p.54;
his Researches on Nitrous Oxide forms volume III of
the Works; the first edition appeared in 1800.
Mitchill went on to become a Congressman.

16. W.Nicholson & A.Carlisle, Nicholson's Journal
(4th series) IV (1801) 179-87; A.Volta, Phil.Trans.,
XC(1800) 403-31,in French; a translation appeared in
Phil.Mag.,VII(1800) 289-311. On journals, see D.M.
Knight, Sources for the History of Science,1660-1914
(1975) ch.4.

17. C.Girtanner, Nicholson's Journal, IV (4th series
1800) 137-40, 167-71, 268-75. See M.P.Crosland, The
Language of Chemistry (1962) pp.207-8.

18. Phil.Mag.,VI(1800)152-3,216-7,335-54, For Winterl's
chemistry, see G.Cuvier, Rapport Historique (Paris,
1810) pp.83-8, and the article on Winterl by M.A.
Snelders, Isis,LXI(1970) 231-40; on Ritter, see W.D.
Wetzels, 'Aspects of Natural Science in German
Romanticism', Studies in Romanticism,X(1971) 44-59;
on the sciences generally in Germany at this period
see the notes by M.Petry to his translation of G.W.
F.Hegel, Philosophy of Nature (3 vols,1971). See
also D.M.Knight, 'Steps towards a dynamical chemistry'

Ambix,XIV(1967) 179-97. On the Dutch chemists, see
T.Levere, 'Relations and Rivalry: Interactions
between Britain and the Netherlands in 18th-century
Science and Technology', History of Science, IX
(1970-publ.1971) 42-51; for the earlier period, see
C.A.Lindeboom, Herman Boerhaave (1968). On national-
ism in science, see M.P.Crosland (ed.) The Emergence
of Science in Western Europe (1975).

19. Phil.Mag.,VII(1800) 76.

20. Mitchill's papers may be found in Phil.Mag.,III
(1799) 177; IV (1799) 35, 42; a report on his later
career appears in Phil.Mag.,XXXVII(1811) 208.

21. W.Whewell, History of the Inductive Sciences
(1837) III, p.19. The passage resembles   Joseph
Priestley's History of Electricity,(1775) I,p.xiii.

22. H.Davy, Consolations in Travel(1830) pp.249-52;
cf his Laboratory Notebook, August 30th 1810; and
Faraday's description in H.Bence Jones Life of
Faraday(1870) I, pp.60ff and 372-3, of his work with
Davy on nitrogen trichloride, when few precautions
were taken. In Chemical Manipulation(1827) Faraday
recommends safer procedures.

23. H.Davy,Works,VIII, p.260. On Davy's electrical
work, see T.H.Levere, Affinity and Matter (Oxford,
1971)chap.2; L.P.Williams, Michael Faraday(1965)
chap.6.

24. On electric fish, see J.Atkins, A Voyage to
Guinea(1735) p.47; H.Cavendish, Electrical Researches
(1879) pp.194-215; A.von Humboldt, Personal Narrative
of Travels to the Equinoctial Regions of America,
trans.T.Ross(3 vols.,1852-3) I, pp.114ff; M.Faraday,
Experimental Researches in Electricity(1839) I, pp.
76-107. W.H.Wollaston, 'Experiments on the chemical
Production and Agency of Electricity', Phil.Mag.,XCI
(1801) 427-34. H.Davy,Nicholson's Journal,IV(1800)340.

25. E.Darwin, The Botanic Garden(1791) p.46 note;
M.Donovan, Essay on Galvanism (Dublin,1816) p.54;
H.Davy,Works,VIII, p.272.

26. J.W.Ritter, 'Die Begrundung der Elektrochemie',
ed. A.Hermann, Ostwalds Klassiker der Exakten
Wissenschaften,neue folge,II,(Frankfurt am Main,1968);
J.Jorpes, Jöns Jacob Berzelius, tr.B.Steele,
(Stockholm,1966); W.Hisinger and J.J.Berzelius,
'Experiences Calvaniques', Annales de Chimie,LI, (30
Messidor An XII) 167-74. Brief versions of Ritter's
papers appeared in Nicholson's Journal,VI(1803) 221-3;
VII(1804) 288-91; VIII(1804) 176-80; see also his
paper in Annales de Chimie,XLI,p.208. On Ritter,

see note 3.

27. C.B.Desormes, Annales de Chimie,XXXVII, p.284; M.Robertson, Annales de Chimie,XXXVII, p.284 — he was followed by a M.Lehot, Annales de Chimie,XXXVIII, p.42; A.F.Fourcroy, N.L.Vauquelin and L.J.Thenard, Annales de Chimie, XXXIX, p.103; J.B.Biot and J.L.N.F.(Georges)Cuvier, Annales de Chimie,XXXIX, p.242; Biot, Annales de Chimie, XLVII, p.1. The journal followed the Revolutionary dating during these years.

28. J.Priestley, Nicholson's Journal,I(1802) 181,198, 202.

29. Sylvester, Nicholson's Journal,IX(1804) 179; 'a correspondent', Phil.Mag.,IX(1801) 352.

30. Governor Pownall, Phil.Mag.,XVIII(1804) 155; for a similar view of the aether, see J.Tyndall, Fragments of Science(1871) p.4; of course Pownall is looking back to such Newtonians as Hartley. See the anonymous article in Phil.Mag.,XVIII(1804) 170; Nicholson's remarks appear in Nicholson's Journal, VIII(1804) 70; G.S.Gibbes elaborated his papers into a book: A phlogistic theory,ingrafted upon M.Fourcroy's Philosophy of Chemistry (Bath,1809); Wilson, Nicholson's Journal,X(1805) 42; H.B.K., Nicholson's Journal,XIII(1806) 40,224; XIV(1806) 50; Thicknesse, Nicholson's Journal,IX(1804) 120.

31. J.Bostock, Nicholson's Journal, II(1802) 296; III (1802) 3,69.

32. H.Davy, Works,VIII, p.272-3.

33. H.Davy, Nicholson's Journal(4° series) IV(1800) 275,340.

34. From the Greek ἀμίαντος undefiled; see the dictionary part of the Encyclopedia Metropolitana, (1845) vol.XIV, p.804, 'asbestos'. Torpedo-fishes gave no shock to females under the influence of nervous fever: W.Kirby, Bridgewater Treatise, ed. T.R.Jones, F.R.S.(1853)II, p.294. A.C.Becquerel employed these cells with two fluids and one metal in some of his studies of electrolysis. H.Davy, Works, II, p.190; Nicholson's remark may be found on p.216; the electric arc is described by Davy in the rare Journal of the Royal Institution,I(1802) 165-7, Works,II, p.212, and in Nicholson's Journal, IV (4°series,1800) 326; see his Syllabus(1802) p.61 for the remarks on galvanism and electricity.

35. J.Bostock, Nicholson's Journal,III(1802) 3,69; On Davy's applied chemistry, see M.Berman, 'The Early

Years of the Royal Institution', Science Studies,II
(1972) 205-40.

36. See J.J.Berzelius, Traité de Chimie (Paris,1829-
33) I,p.165, and Essay sur la théorie des proportions
chimiques,(Paris,1819),reprint,introduced by C.A.
Russell,1972; F.T.Grotthus, Phil.Mag.,XXV(1806) 330-9.
M.Donovan, Essay on Galvanism (Dublin,1816) p.87.

37. H.Davy, 'Bakerian Lecture', Phil.Trans.,XC(1807)
1-56; the lecture described 'some chemical Agencies
of Electricity'. The ninth edition of the
Encyclopedia Britannica (Edinburgh,1875) article
'Glass' gives compositions for various kinds of glass
rather later in the century. Davy's first experiments
on water may be found in his Works,II, p.206;
Sylvester, in Nicholson's Journal,XV(1806) 50, was
among those who detected products other than oxygen
and hydrogen.

38. The reasoning sounds cool and inductive, but the
notebook is as usual a mess; Professor R.Siegfried
has spent much time with these fascinating MSS, of
which an edition is a desideratum. The passages
quoted come from 30 October 1806, and from 11 November
1806. See T.Levere's papers on Davy and Coleridge,
Studies in Romanticism,XVI(1977)349-79, and Annals of
Science,XXXV(1978)33-44.

39. H.Davy, Phil.Trans., XCVII(1807) 12. Among those
who doubted was J.B.Stallo; The Philosophy of Nature
(1848) p.91, for whom elements were formed during
electrolysis and were not pre-existent.

40. T.Thomson, The History of Chemistry, 2nd edn.
(1830-1)II, p.260; see A.Kent, 'Thomas Thomson:
Historian of Chemistry', British Journal for the
History of Science,II(1964) 59-63. Outbursts against
Davy's lecture,by M.D. and by R.Harrington, appear
in The Gentleman's Magazine, LXXVIII(1808) 600,694;
Harrington, who had written proving the Sun and
planets to be inhabited, published in 1804 The Death
Warrant of the French Theory of Chemistry, and in
1819 a very necessary, but unsuccessful Elucidation
of the Harringtonian System of Chemistry.

41. J.Priestley, History of Electricity(1775) I,
pp.xiv-xv. J.J.Berzelius, Traité de Chimie (Paris,
1829-33) I, p.164; The Edinburgh Review XII(1808) 394.

3

Chemistry and Materialism

The history in the nineteenth century of the belief
that all matter is ultimately of one kind — or at
most two or three — seems, as we saw in the introduction, open to analysis in terms of Comte's[1] three
stages theological, metaphysical, and positive. In
the early years of the century, investigations by
Davy and others of the role of active principles or
agencies in chemistry seemed to overturn the
materialistic, and anti-Newtonian, doctrine that the
properties of a substance depended upon the properties
of its constituents, and that matter was inherently
active or possessed of powers. Later investigators,
filled with the doctrine of the simplicity and
harmony of nature, pursued Newton's conjectures on
the possibility of converting ordinary matter, whatever body it happened to compose, into light; or
impressed with the success of the organic chemists in
erecting their edifice upon 'radicals' (composed in
the main of the four elements hydrogen, oxygen,
nitrogen, and carbon) sought to demonstrate that the
elements of inorganic chemistry were not irreducible
but were themselves radicals. While their arguments
were impressive, a complete lack of success attended
all their efforts at transmutation; and it was not
until the fields of spectroscopy, cathode-ray
investigations, and radioactivity were opened up that

the belief in the unity of matter, widely held
throughout the century, began to seem clearly
justified by the facts.

In this and succeeding chapters, we shall
follow this artificial but convenient schema, tracing
the belief from its theological stage through its
metaphysical to its positive stage; remembering that
the 'stages' are not really distinct, and that we are
dealing with the chemistry and physics throughout.
The different views therefore had some empirical basis
at each point in the story. We shall also have to go
back into the eighteenth century at intervals; for
the various theories and attitudes have histories, of
which their nineteenth-century adherents may or may
not have been aware, and terms like energy and power
had gradually acquired their then current meanings,[2]
though these definitions were still unstable. Neither
the men of science, in short, nor their terms had
been born yesterday; and even if a man used a term in
a new way, his contemporaries would be expecting him
to use it in the way they understood. They would also
expect him — and usually rightly — to be attempting
to answer questions of the kind which they and their
predecessors asked.

Davy confirmed Lavoisier's view that water was a
compound of oxygen and hydrogen, and that the alkalies
and earths were metallic oxides. But his work can be
seen as a reaction to, as much as a development of,
that of Lavoisier. His first paper was directed
against the caloric fluid, and the idea that changes
of state were to be interpreted as chemical reactions
of solids and liquids with this fluid. Later, as
Lavoisier had urged chemists not to do, Davy did
speculate about the ultimate particles of bodies
instead of resting content with elements; and he

toyed publicly with the theory of phlogiston. He
was delighted to prove that Lavoisier had erred in
choosing the name 'oxygen', since the strong acid
hydrogen chloride did not contain any of this
'generator of acids'. Davy was one of a generation
to whom the French seemed characterised by materialism and infidelity[3] (and consequently by aggressive-
ness); and as Coleridge turned to the divines of the
seventeenth century to get away from the idea of God
as First Cause only, so Davy turned to the Newtonian
tradition of corpuscles and active powers in place of
principles or elements which were the bearers of
certain properties.

In this last tradition, we find the physician
William Heberden[4] seeking for the principle of life;
phlogiston was the principle of inflammability; and
oxygen in the new chemistry was the principle of
acidity. All these were usually considered as
material constituents without which a body could not
have the property of which they were the bearer.
Davy himself sought at one time for the principle of
metallization, and supposed that the analogy in the
properties of potassium and ammonium salts indicated
that potassium and ammonium had a common constituent
— probably hydrogen. Like Faraday, Davy displayed a
capacity for the creative misuse of apparatus; he
worked fast, adapting what was available to his needs
before bystanders realised what he was doing,[5] and it
may be that he displayed a similar opportunism towards
theories. However if we follow opposition to materialism as our thread, we may find our way through the
maze-like writings of Davy and his contemporaries with
less difficulty. In terms of this analysis, Davy
showed first that substances were not fluids or gases
because they were compounds of the supposed element

caloric; and then that acidic properties were the result not of combination with oxygen but of certain arrangements of matter under active powers.

A different version of materialism — without the property-bearing elements — was that of Priestley,[6] who interpreted his contemporary Boscovich's atomic theory to mean that matter was possessed of powers, which were indeed all that we ever encountered. One could not touch the physical points which for Boscovich constituted the atoms, for one met only the associated repulsive force rising towards infinity very close to the point centre. For Priestley, if matter were active then there would be no need for any immaterial substance; and following Newton's Rules of Reasoning we should reject it. This suited Priestley's theology, for he believed in the Resurrection of the Body and wished to throw out the Immortality of the Soul along with that other 'Platonic corruption', the doctrine of the Trinity. Davy, on the other hand,[7] was converted to immaterialism in his late 'teens, and seems to have adhered to a scheme including brute matter and immortal souls, with light, electricity, and magnetism coming somewhere in between: usually as powers distinct from matter, but sometimes as refined aethereal matter. Such experiments as changing the chemical reactivity of a body by giving it a positive or negative charge — described in Davy's Bakerian Lecture of 1806, and used in his abortive researches on cathodic protection for copper-bottomed ships in 1823-6 — find a ready explanation in terms of powers separate from brute matter, though like any experiment they do not entail any one explanation. If the chemical properties of a piece of iron or copper depend on its electrical charge, then they cannot be inherent in its constituent particles.

To settle exactly what Davy believed at a given
time would require careful study of his manuscripts;
and even this could not put the question beyond doubt.
We, dealing here with public science and therefore
mainly looking at published material, must judge as
did his contemporaries, especially those who were
not intimate with him, and see what use was made of
his discoveries. That it was possible to see them,
and those of others, as armour against materialism is
clear from what Coleridge wrote[8] in his introduction
to the Encyclopedia Metropolitana in 1818. After
praising those who, in discussing electricity, had
avoided a 'notional fluid or fluids, ... chemical
compounds, ... [or] elementary matter' in favour of
'the idea of two-opposite-forces, tending to rest by
equilibrium'. Coleridge accounted for the charm of
chemistry by suggesting that the science 'is a pursuit
of unity of principle, through a diversity of forms'.
And in the Conclusion to Aids to Reflection, he
remarked how the dynamical science of Davy, Oersted,
and John Hunter was ousting Cartesian dualism and
mechanism. It would be possible to construct a
treatise on the eye from the materials which compose
eyes; when we speak of an eye, therefore, we are not
referring to its 'sensible and ponderable Stuffs or
elements'. The term includes the 'Principle, the
Operating Cause of the Combination'; if it did not,
we should not be able to distinguish a living from a
dead eye. A dead eye is only an eye as a heap of
ruins in the desert is Babylon or Persepolis.

Just as a person is an integrated whole of body
and spirit, not a ghost in a machine, so a chemical
compound is a genuine tertium quid and not a mere
juxtaposition of particles. The distinction between
juxtaposition and synthesis made by Coleridge and

65

Others at this time had a long history;[9] it was made
by the Stoics in their discussions of the union of
body and soul, and is found in Augustine. Davy
discovered the very different properties of the
various oxides of nitrogen — from air, which seemed
a weakly-combined compound, to the delightful
laughing-gas nitrous oxide, and then to the brown,
choking nitrogen peroxide — which all have the same
two components; and proved the chemical identity of
charcoal and diamond, which have such different
physical properties. Both facts were compatible with
a view of chemical combination as a synthesis, but
incompatible with the doctrine that chemical composition is merely an affair of the juxtaposition of
particles.

It is then worthwhile taking a closer look at
Davy's work in this light. Like Goethe, he was proud
of being inconsistent, and we shall not find orderly
progress, or formal statements of his views.[10] In
his famous introductory lecture of 1802 he declared
that chemistry still lacked

> a precise and beautiful theory.
> Though we can perceive, develope,
> and even produce, by means of our
> instruments of experiment, an
> almost infinite variety of minute
> phenomena, yet we are incapable
> of determining the general laws
> by which they are governed, and
> in attempting to define them, we
> are lost in obscure, though sublime
> imaginations concerning unknown
> agencies. That they may be dis-
> covered, however, there is every
> reason to believe.

The urge to investigate the phenomena of 'corpuscular
action' was natural and instinctive; for the bored
city-dweller chemistry was exhilarating because it

exhibited 'an almost infinite variety of effects
connected with a simplicity of design.'

In the Syllabus of Lectures for 1804[11] on
Chemistry in its connexion with natural operations,
we find the heading 'Of the Elementary Particles, and
the active Power of Matter'. Of the elements, Davy
wrote that

> in their natural arrangements they
> are acted on by attraction: and to
> the agencies of this power, as
> chemical affinity; and to the
> effects of heat, light, and
> electricity, produced by imponderable
> substances, as yet but little
> understood, almost all the
> corpuscular changes of dead matter,
> either immediately or remotely
> belong.

Heat, light, and electricity seemed possessed of
certain 'agencies'; thus heat can call into a state
of activity elements apparently inert. The phenomena
of electricity seemed to be due to a universally-
diffused medium, put into an active state; experiments
made on the discharge of electricity through vacua in
the late eighteenth century and later by Davy were
designed to investigate this view.

Davy's first Bakerian Lecture of 1806 introduced
polarity into the mainstream of chemistry, and thus
maintained among chemists that interest in forces
rather than weights which Proust and Dalton seemed
likely to kill.[12] Chemistry could still be described
as research into the powers which modify matter. The
idea that chemical affinity and electricity were
manifestations of one power was to prove extremely
fertile; and to raise anew the problem of the
relationship between matter and the powers. In the
lecture, the language of polarity was not directly

used; we find instead such terms as 'repellant and
attractive energies', 'powers', 'the balance of
natural electrical energies', 'the balance of power',
and 'positive and negative electrical energy'. The
words 'energy' and 'power' in the early nineteenth
century carried anthropomorphic connotations;
'energy' was used of living, rather than inanimate
things, and 'power' of cosmic forces like St Paul's
'principalities and powers', or of human potentiality,
rather than in mechanics, as in 'horse-power'. We
shall explore this later.

In January 1807 Davy began his lectures with
'the consideration of the active and etherial powers,
as they may justly be called, which are continually
operating upon gross and ponderable matter.'[13] In
his last formal course at the Royal Institution, in
1812, he lectured on the great forms — that is,
states — of matter, and the active powers of nature,
which were:

> Those powers which cannot be
> separated from the bodies which
> they affect, and which produce
> the motions of their particles,
> ... as gravitative, chemical, or
> as electrical attraction. Active
> powers must be considered as be-
> longing to matter; but it is not
> necessary to suppose them inherent
> in it. It may be regarded itself
> as inert; and all effects produced
> upon it as flowing from the same
> original cause, which, as it is
> intelligent, must be divine. ...
> The different powers in nature
> must be referred to one source.

There follows the suggestion that whereas God was the
architect of the Universe, the powers were its
builders.

The first of these quotations came from a lecture

68

only a matter of weeks after the Bakerian Lecture of 1806, and the second came some time after his work on chlorine and on acids. But between these two came a period of uncertainty. In 1808 in another lecture, Davy remarked that:

> Change is the essence of all the operations in nature, and change may constantly be referred to motion. But for motion there must be a moving cause, and this necessarily implies the existence of active powers. ... By the tendencies of substances, to approach or recede from each other, the various alterations of the forms of things, whether obvious or obscure, rapid or slow, may be explained, and attraction and repulsion established as motive agents or influences. In our discussion on every part of the science of nature, we are obliged to revert to these powers. But though sound philosophy permits us to allow them as active, it is far from considering them as inherent in or necessarily attached to matter. After the example of Newton, they may be considered as principle, but not ultimate, as secondary, and not primary causes. In nature nothing can be said to be automatic; — one order of events flows from another, an immense number of phenomena depend upon a single law, but all may be considered as a work of mechanism, and of divine institution. In human inventions, springs may move springs, and wheels wheels, but the first motion, and the regulation, must be derived from the artist. ... The phenomena of electrical attraction and repulsion ... are not merely important in themselves, but they are guides in our future progress in electro-chemical science ...[14]

Davy, in this Newtonian vein, thought it probable
that every species of attraction and repulsion,
celestial and terrestrial, might be referred in the
future to one grand and universally-operating law;
although it was still necessary at present to
distinguish the various species.

But even at the period of this Newtonian
rhetoric, Davy was turning away from this view of
brute matter and active powers towards a more chemical
and materialistic conception. This in fact led him
into the mire from which he only emerged in 1810 with
the new view of chlorine and of acidity. If the
Newtonian view of matter was sterile in chemistry in
the late eighteenth century, in the early nineteenth
it displayed a renewed fecundity.[15] Davy's Bakerian
Lecture of 1807 did indeed describe the isolation of
potassium and sodium, in accordance with his views of
1806, and it was suitably greeted with enthusiasm in
the learned reviews of the day. The properties of
these new substances — particularly their low density,
for heaviness was held to be a metallic characteristic — aroused great interest, and led to their becoming important analytical tools for chemists. But
Davy began in this very lecture his series of
phlogistic speculations. He realised that his own
discoveries had shaken the antiphlogistic system of
Lavoisier, which might be overturned by 'the
knowledge of the powers of the etherial substances'.
Davy said, as Beddoes had of the artificial but
convenient Linnean classification of plants, that he
was prepared to use Lavoisier's system for its beauty
and precision, but did not feel confident of its
truth and permanence. In its place he advanced the
materialistic view that all the metals were 'compounds
of certain unknown bases with the same matter as that

existing in hydrogene; and the metallic oxides, alkalies and acids compounds of the same bases with water'. He conceded that this theory would be less elegant and distinct than that of Lavoisier.

Davy supposed all metals to contain hydrogen, but not in the way in which ordinary compounds of that element do; for he continued to oppose the view that potassium and sodium were hydrides, believing that the status of these metals was the same as that of all the others.[16] When Berzelius discovered 'ammonium amalgam' on electrolysing an ammonium salt with a negative pole of mercury, the evidence for hydrogen in all metals, or for the metallic nature of hydrogen or nitrogen, was increased; for the analogy between the compound radical ammonium and the metal potassium was shown to be even closer. Davy wrote that since all combustible bodies and metals went to the negative pole, it was probable that all of them contained hydrogen; and that similarly oxygen was the only negative elementary body, so that everything which went to the positive pole must contain it. The 'phlogiston or hydrogene' of platinum was most, and that of ammonium least, tightly combined. In researches reported in the Bakerian Lecture for 1808, Davy seemed to have isolated hydrogen from phosphorus and sulphur; he believed that if these substances could be obtained free from oxygen, they would be metallic.

The objective here was to bring about a reduction in the number of the basic entities of chemistry; to produce 'a new, a simple, and a grand philosophy'. If there be, said Davy in a lecture in 1809, 'some one principle of inflammability, then strict analogy would lead us to conclude that there must also be a

metallizing principle, some substance common to all
metals'.[17] But in fact it was when the search for
this principle or substance was given up in favour
of the doctrine of powers and arrangements that Davy
solved the enigma of muriatic acid; and it is odd
that the passage quoted goes on to hope for a
mathematical chemistry of short-range forces in which
the 'electrical energies' of bodies, corresponding
to their chemical attractions, would be measured and
compared. Indeed Davy often at this period gave in
lectures and papers alternative accounts, in terms of
principles and of powers, of how the phenomena might
best be explained. Thus his discussion of the
similarities of potassium and ammonium contains the
remark: 'It is not impossible that matter of the same
kind, possessed of different electrical powers, may
exhibit different chemical forms';[18] and in the
'phlogistic' lecture of 1809 he referred to the
differences between the oxides of nitrogen, and to
the behaviour of some of them as radicals, as evidence
that 'substances which we at present conceive to consist of different species of matter may ultimately be
referred to different proportions of similar species'.

In his first paper of 1799 Davy argued that
light must be a component part of oxygen gas since it
is emitted in so many reactions of oxygen;[19] and an
echo of this is found in his posthumously published
Consolations. But his mature view in 1810 was that:
'The vivid inflammation of bodies, in oxymuriatic
acid gas, at first view appears a reason why oxygene
should be admitted in it; but heat and light are
merely results of the intense agency of combination.'[2]
Though the effect of Davy's work on chlorine — which
we shall examine in detail in a later chapter — was
to throw some into trepidation, it was welcomed by

others who had been unhappy about the materialism of
the French theory.[21] Thus a reviewer in the
Philosophical Magazine applauded Davy's vitalism and
his opposition to materialism, and wrote that his
was 'a much more simple and natural exposition of
the actions of matter, than the supposed legerdemain
operations of phlogiston or oxygen'. Similarly, John
Mason Good in lectures at the Surrey Institution,
declared that only to the corpuscular theory did the
discoveries of modern chemistry give countenance.
The supposed elements, earth, air, fire, and water,
had given way to others like oxygen; but these were
now tottering, and all would fall as 'the Proteus-
form of matter' was followed into its deepest
recesses.

Vitalism is a doctrine difficult to define; but
Davy seems to have believed that the phenomena of life
were not simply the outcome of various arrangements
of matter, but involved new powers. In 1801 Coleridge
had written to him[22] that 'all power of vital
attributes' depended upon arrangements, which
chemistry did not yet illuminate. Good attributed to
Davy the view of Aristoxenus that 'Life is the result
of a general harmony or consent of action between the
different organs of which the vital frame consists.'
But Davy seems further from materialism than this
implies; thus in 1808 he declared that the powers
concerned in life must be for ever hidden: 'that
which sees, will not be visible; — that which touches,
will not be felt; — that which commands sensations,
will not be their subject.' In his Elements of 1812
Davy suggested — invoking the names of Hooke, Newton,
and Boscovich — that different arrangements of the
same elementary particles might constitute all the
bodies in nature. Liebig was to suggest that

chemical forces were not altogether dissimilar to vital forces and that "the active force in a compound depends on a certain order or arrangement"; this force leading for example to acidity or alkalinity.[23] Davy was not happy about this parallel between the vital and the chemical; and seems to have been more interested in the powers than in the arrangements.

Davy[24] did take up the suggestion found in the works of the seventeenth-century theologian Ralph Cudworth that if matter be indestructible, then a fortiori the soul must be immortal; writing about 1817:

> If matter cannot be destroy'd
> The living mind can never die;
> If e'en creative when alloy'd
> How sure its immortality!

In his Consolations in Travel Davy elaborated his views of the relation of science to a general world-view. Like Boethius, Davy had been removed from high office in middle life and was under sentence of death; though in Davy's case the cause was not imperial wrath but disease. Davy was seeking consolation for this and more generally for the decay of his power of imagination, which had enabled him to discern true rather than fanciful analogies in nature. His message was that as the soul goes on developing it outgrows its material machinery; and that death is to be welcomed as a rebirth in a more-aetherial body. The prospect is of an indefinite series of rebirths for Davy believed — with Origen — that this process will go on until all spiritual natures have passed through every mode of life in their progress towards power and knowledge. He implies in one dialogue that the relationship of God to the world is the same as that of our minds to our bodies, except that the world does not run down. He distrusted the use of 'evidences' and of the unaided reason in religion; we should not speculate, but should submit to God's inscrutable will, as animals follow their instincts. Religion is indeed a 'pure and divine instinct', and Revelation supplies the place of ordinary instincts in man.

The dialogue of most interest to us is the fourth, 'The Proteus'.[25] We are familiar with Proteus who would answer riddles if held through all his changes of shape, and the idea of matter as Proteus was a commonplace. The creature Proteus Anguinus was described in the Philosophical Transactions for 1801 as an ambiguous amphibious creature; living in caves, it had eyes, but seemed blind; its origin was mysterious — perhaps it was flung from a subterranean sea — and it was unclear whether the animal was perfect in itself or was the larva of a higher species. It was therefore in all points a symbol of man, the 'Great Amphibium'. The dialogue is concerned with the possibility of a materialistic biology. The Unknown, who represents Davy, said that he did not 'consider vitality dependent upon any material cause or principle'. To suggest that matter simply by organisation acquired powers not naturally possessed by it was unphilosophical; no juxtaposition, refinement, or arrangement of the particles of matter could give rise to intelligence, and God could not have bestowed the power of thought on matter any more than he could have made a house its own tenant. Life therefore depends upon an immaterial principle; theories of the vital heat have not explained the divine fire. The imponderable agents heat, light, electricity, and

75

magnetism, were now described as 'refined forms or modes of existence'; souls arrive wrapped in some such aetherial matter, to which, according to Davy, is to be attributed instinct and conscience which could not inhere in immaterial substance.

The progress of chemistry led through the work of Liebig and then of Kekulé to a tougher notion of the arrangement of particles[26] — Daltonian property-bearing atoms rather than Newtonian corpuscles — culminating in the 1860s in atomic models made of balls and wires which Brodie was to denounce as 'materialistic joiner's work'. Meanwhile the development of the Principle of Conservation of Energy in the 1840s provided a new way of describing the imponderable agents. The dichotomy of body and spirit ceased to offer parallels to those concerned with the relations between matter and energy. In view of this, it is surprising to find that Davy's Consolations went on selling into the middle of the nineteenth century. A French translation appeared in 1869, apparently with a ninth edition in 1883; and the work was translated into German both in whole and in part, the latter having — significantly for our purposes — the title Goldkörner aus dem literarischen Nachlasse eines christlichen Naturforschers. Among specifically-theological writings in English, one finds references to a physics of powers in Coleridge and in F.D.Maurice; but it could not be claimed that this chemical contribution was a very important or lasting part of Victorian theology. It was even double-edged, for the materialistic author of Vestiges described electricity as being as metaphysical as ever mind was supposed to be, in the hope of reducing the latter to the realm of cause and effect.

We do find powers in some later scientific

literature; for example in J.P.Nichol's <u>Architecture</u> <u>of the Heavens</u>.[27] Nichol was a former clergyman, who introduced Wordsworth to the poetry of Longfellow, whose books interested George Eliot in science, and whose lectures gave William Thomson an impulse towards physics. Nichol wrote that: 'the period has happily long gone by, in which men whose instincts were at the mercy of any trick of logic, could obtain assentient hearers for the doctrine, that such powers are mechanical agencies unoriginated and uncontroll- ed.' He believed that the reality of power and force implied free will; and seems to have toppled over into pantheism — always a risk in this kind of theology — in asserting that the forces of nature were not powers delegated by God but that 'above, below, around — <u>there</u> is God.' William Prout argued in contrast to this, in his Gulstonian Lecture for 1831 before the Royal College of Physicians, that the organic agent 'must be considered as an ultimate principle, endowed by the Creator with a faculty little short of intelligence, by means of which it is enabled to construct such a mechanism from natural elements, and with the aid of natural agencies, as to render it capable of taking further advantage of their properties, and of making them subservient to its use.'[28] This view of nature would not, in Prout's view, lead to materialism but rather raise our view of the Deity, for it is 'more consonant to our notions of infinite wisdom and power' to suppose that God created 'agents and materials originally endowed with all the energies and properties we have assigned to them' than to credit him with the performance of perpetual miracles.

With Nichol's pantheism and Prout's view of nature as demiurge, we seem to be far from the nineteenth century; and the same impression is derived from reading the Bridgewater Treatise by the elderly clerical entomologist William Kirby.[29] The Bridgewater Treatises are a curious collection of works on natural theology, which it is difficult to take as seriously as one should as an historian; and Kirby's is undoubtedly one of the most curious. He had with William Spence written the standard English work on entomology, which runs to four volumes and has an impressive bibliography; and the part of the Bridgewater Treatise which deals with the natural history of the invertebrates is sound enough. It is in his general discussions, particularly of those 'great powers that, under God, rule the physical universe', that the modern reader feels at sea. We are not surprised to find that Kirby was a friend of the anti-Newtonian William Jones of Nayland, and was sympathetic to the ultra-pious Hutchinsonians. He believed that Lamarck had been led to propose his theory of evolution — which no one in his rational senses could believe for one moment — because of his materialism. The sensible course would be 'for the philosopher to ascertain what has really been revealed in Scripture on these great subjects.'

Naturally we do not find this out merely by taking the texts at face-value, for the truths therein were concealed, from the ignorant who could not bear them, beneath a veil of allegory. The Scriptures must therefore be decoded; and Kirby's achievement was to identify the cherubim, upon which God is said to sit and ride, with the powers in nature.[30] The term Heaven also applied properly to the powers which maintained the motions of the celestial and terrestial spheres, 'all derived originally and still maintained, from and by one parent fountain, the centre of all

78

irradiation, of all light, of all life and energy'.
Two cherubim looking towards each other at either
side of the Mercy-Seat represent the antagonist
powers, vis centrifuga and vis centripeta; these two
form the pillar of the world, universal gravitation,
and though they may be called 'antagonist or conflict-
ing powers, yet their opposition is not enmity, but
universal harmony and love'. Philo is called in
support of this view; and the analogy of man the
microcosm and the great world or macrocosm indicated
to Kirby that as in man there is a continual flux
and reflux of blood, so there must be cyclic processes in the heavens maintaining the radiation from the
sun. Secondary causes and mechanisms were not
sufficient of themselves 'to keep the whole machine
together and at work'; we must see God everywhere as
the mainspring of the universe, which He governs
through the cherubim or powers. These powers were
probably also responsible for the physical development of animals, and for their instincts; which
Addison had indeed compared to gravity in inanimate
matter.

Clearly the sense in which Kirby used the term
'power' was very different from that of his more
'modern' contemporaries[31] such as William Hyde
Wollaston the chemist, and John Farey the engineer.
It is hard to say on which side Davy, that avid
reader of the Arabian Nights, would be found; all
that we need to remark is that Davy's Consolations
was drawn upon by Kirby, who in his discussion of
instinct compared Davy's Ethereal Matter with the
Spirit of Nature of the Cambridge Platonist Henry
More. Kirby was clearly neither a Romantic nor a
devotee of Naturphilosophie; his curious discussion
of the powers therefore indicates that it was

possible to believe in a world in which force or power was paramount, in which the opposition of forces led to a higher synthesis, and in which there was an underlying unity of the various powers. These are doctrines which one associates with Schelling, Ritter, and Oken; but as in the biological realm similarity of form need not imply community of descent, so it is not necessary without positive evidence to invoke direct influences here. Even Coleridge came to Schelling after reading Proclus and Plotinus and their English disciples, as well as David Hartley, Priestley, and the Philosophical Transactions: and for scientists there is no need to invoke a foreign tradition if we can demonstrate that there was a native one. The existence in eighteenth-century Britain of a tradition that atoms were centres of force has been demonstrated by Arnold Thackray and Robert Schofield, and it can indeed be described as orthodox Newtonianism. Our task will be briefly to trace this view down to the time of Davy, and to indicate how it could seem that in opposing French materialism Davy was employing new experiments and discoveries but an old view of the world, as much as anything novel coming from Germany.

Kirby coupled the names of Davy and More; and it is to the writings of the Cambridge Platonists, a group which included More, Cudworth, the Royal Society's publicist Joseph Glanvill, and John Ray the great naturalist, that we should first turn.[32] In More's poetry we do find the suggestion that the soul is clothed with aether, the vehicle of sense; and that matter is inert, God being the source of all vitality. There is a more-extended discussion of spirit and matter by him in Glanvill's book of ghost-stories, Saducismus Triumphatus; in which he makes

the distinction between incorporeal and imponderable
substance. More and Cudworth were both atomists;
their atoms were inert and massy, and were moved and
arranged by powers. The property-bearing atoms of
the homoimeria of Anaxagoras were rejected by
Cudworth, along with materialism and fatalism; he
attributed the atomic theory to Moses, and urged that
the corruptions introduced into it by such atheists
as Democritus be abandoned. The indestructibility
of matter proved only that souls were immortal.

From the Cambridge Platonists emerged in the
next generation the latitude-men, liberal but
orthodox theologians with a particular interest in
natural theology, who seem to have been much more
influential in eighteenth-century England than such
deists as Collins, Tindal, and Toland. William
Wollaston,[33] the grandfather of the chemist, wrote a
famous work of this kind, The Religion of Nature
Delineated, which has a certain Newtonian significance
in that Benjamin Franklin worked as a compositor on
the edition of 1725, when he was in London. We find
in this book a naturalistic ethic, a calculus of
pains and pleasures, and a scale of nature including
incorporeal beings; and then an argument to show that
matter cannot think, and that the soul cannot be
material: materialism must be false; matter is inert
and passive and could not be made to think without
becoming another kind of substance. The philosophical ships-surgeon John Atkins similarly considered it
impossible to explicate the cohesion of matter and
the diverse properties of different bodies 'without
putting the Deity to it'; and Maclaurin, in his
standard account of Newton's discoveries, described
the powers — which might receive an explanation in
terms of an aether — as God's instruments rather than

81

His immediate volitions.

The most famous development of Newton's remarks about the aether was that of David Hartley,[34] who made thoughts into vibrations in the aether, associated together to form complex ideas. This scheme sounds materialistic, and indeed Priestley was to emphasise this aspect of it; but Hartley himself devoted much space to proving that his views did not weaken natural religion, and argued that matter was 'a mere passive Thing, of whose very Essence it is, to be endued with a Vis inertiae'. Priestley did take the further step of making matter active and spirit or immaterial substance redundant. His advocacy, in the Disquisitions on Matter and Spirit, of point atoms which were centres of force made Boscovich's atomism well-known in Britain by the end of the century; but his materialism seems to have been regarded with the kind of horror with which earlier men of science had contemplated the doctrines of Hobbes. Even in the 1860s it was possible to use as a major objection to an atomic theory that it was materialistic; half a century earlier this argument would have been very powerful.

Among the critics of Hartley was the Scottish 'common-sense' philosopher Thomas Reid, who express-ed his disapproval of the manner in which after 'expressing his approbation of the proper method of philosophizing, pointed out by Bacon and Newton', Hartley should have proposed a system 'built partly on facts, and much upon conjecture'. For Reid, Hartley's system had not surprisingly led to false conclusions being drawn. Philosophers had anyway mistakenly deserted the common-sense view that we perceive objects in favour of the view that we perceive images; this had led to a misunderstanding

of the role of organs. The eye does not see any
more than a telescope does; it is merely an instrument through which we see. The 'impressions of
external objects upon the machine of our bodies'
cannot be the real efficient cause of thought and
perception; the mere suggestion is absurd. Reid
forcibly made the point that mind and body are in
different categories; there appears to be a vast
interval between them, and we know of no inter-
mediate nature that connects them together:

> We have no reason to ascribe
> intelligence, or even sensation,
> to plants; yet there appears in
> them an active force and energy
> which cannot be the result of any
> arrangement or combination of
> inert matter. The same thing may
> be said of those powers by which
> animals are nourished and grow,
> by which matter gravitates, by
> which magnetical and electrical
> bodies attract and repel each
> other, and by which the parts
> of solid bodies cohere.

The continual operation of intelligent beings may be
required to maintain these powers; or there may be
beings active without intelligence, 'a kind of
incorporeal machinery.'[35]

We know that Davy had read in his 'teens some
of the writings of the common-sense school, and there
are echoes of Reid in the Consolations.[36] But all
that we need to know is that this view of chemistry
survived into the nineteenth century; so that talk
of active powers would be well-understood, and
opposition to materialism was a real force. Davy's
argument that heat was not a material substance; his
demonstration first that the alkalies contained
oxygen, so that this substance was not simply an acid-

generator, and second that muriatic acid contained
no oxygen, so that acidity was not produced by the
presence or absence of a material component; and then,
when he had married a wealthy widow and could afford
to burn a diamond, his experimental proof that
diamond was chemically identical with charcoal, so
that physical properties like transparency and hardness did not depend upon a material component either;
all seemed to confirm the Newtonian dynamical view
of the world. Davy himself, as we have seen, sometimes tried other views; but this was the one which
brought him success. Materialism remained a focus
of interest among men of science in Britain, but the
emphases shifted from chemical data as the century
wore on;[37] we may remember Tyndall's notorious Belfast
Address, and the activities of the Society for
Physical Research. We shall not pursue it into these
regions, but look at another Newtonian survival, one
which lasted through the nineteenth century; the
belief that matter might be put into a fourth or
radiant state, in which its particles would consist
of very simple arrangements, of very few kinds, of
the basic corpuscles of matter. We thus leave the
theological for the metaphysical stage of the belief
in the unity of matter: setting the stage with a
speculative paper from the later nineteenth century.

NOTES

1. A.Comte, The Positive Philosophy, 1853.

2. D.S.L.Cardwell, 'Early Development of the
Concepts of Power, Work and Energy', British Journal
for the History of Science, III(1867) 209-24. See
also R.E.Schofield, Mechanism and Materialism
(Princeton,N.J.,1970); A.Thackray, Atoms and Powers
(Cambridge,Mass.,1970).

3. T.L.Hankins, Jean d'Alembert (Oxford,1970) p.98;
see D.M.Knight, 'Chemistry, Physiology, and
Materialism in the Romantic Period', Durham University
Journal,LXIV(1972) 139-45; H.W.Piper, The Active
Universe (1962); T.McFarland, Coleridge and the
Pantheist Tradition (Oxford,1969); J.B.Morrell,
'Professors Robinson and Playfair and the Theophobia
Gallica: Natural Philosophy, Religion and Politics
in Edinburgh,1789-1815', Notes and Records of the
Royal Society,XXVI(1971) 43-63; J.G.Burke, 'Kirk and
Causality in Edinburgh,1805',Isis,LXI(1970) 340-54.

4. W.Heberden, Commentaries on the History and Cure
of Diseases (1802) pp.482-3; Heberden was an uncle
of William Hyde Wollaston. Cf the astronomer Chappe
d'Auteroche, A Journey into Siberia (1770) pp.322-3.

5. J.Davy, Memoirs of H.Davy (1836) I,pp.254-6;
M.Faraday, Chemical Manipulation (1827) is full of
recommendations on how broken bits of apparatus may
be turned to new uses — a reprint of this has been
published (1974) ed.G.Porter

6. J.Priestley, Disquisitions (1777) especially
sections I and II; R.E.Schofield, A Scientific
Autobiography of Joseph Priestley (Cambridge,Mass.,
1966) pp.166-71 for Priestley's exchange with
Boscovich; R.Boscovich, A Theory of Natural
Philosophy, tr.J.M.Child (Cambridge,Mass.,1966).

7. J.Davy, Memoirs of H.Davy (1836) I,p.26 — see
also p.121. On p.119 Davy fashionably described
himself, in pantheistic vein, as having had 'a
distinct sympathy with nature' when lying on a rock
on a windy day; there are echoes of pantheism in his
Consolations (1830).

8. Encyclopedia Metropolitana (1817-45) I,p.11; the
text is edited, and the whole preface in its original
form may be found in S.T.Coleridge, The Friend, ed.
B.S.Rooke (1969) I,pp.448-524; especially 467-71, and
p.479 — ignore footnote 1 on the latter page, the
reference being not to Volta or Laplace but to Ritter
and Schelling. Coleridge, Aids to Reflection, ed.
T.Fenby (Liverpool,1874) p.355 (Conclusion, S.10).
The reference to Babylon was topical; C.J.Rich,
Memoir on the Ruins of Babylon (1815). Cf W.Wollaston,
The Religion of Nature Delineated (1726) p.127;
Benjamin Franklin worked as a compositor on the 1725
edition of this famous book — the text was reset in
1726 but the pagination is the same. William
Wollaston was the great-grandfather of William Hyde
Wollaston.

9. A.H.Armstrong (ed.), The Cambridge History of Later Greek and Early Medieval Philosophy (Cambridge,1967) p.358; the Greek terms are παράθεσις and κρᾶσις ; R.Cudworth, The True Intellectual System of the Universe (1678) p.833; R.Siegfried, 'Sir Humphry Davy ... on the Diamond', Isis,LVII(1966) 325-35.

10. D.Luke and R.Pick (eds.), Goethe:Conversations and Encounters (1966) pp.218-9; J.Davy, Memoirs of H.Davy (1836) I,pp.26, 121; on the lecture, see R.Sharrock, 'The Chemist and the Poet', Notes and Records of the Royal Society, XVII(1962) 57-76; H.Davy, Works,II,pp.320, 325.

11. H.Davy, Works,II,p.442; cf pp.445, 446. For experiments on electricity in vacuo, see the next chapter: H.Davy, Phil.Trans.,CXII(1822) 64-75.

12. H.Davy, 'Bakerian Lecture', Phil.Trans.,XCVII (1807) 1-56. See for example 'power' in the dictionary volumes of the Encyclopedia Metropolitana (1817-45)XXIII,p.519. Dynamics went into eclipse in chemistry; cf the section on manipulation in Lavoisier's Elements of Chemistry(1790), which describes calorimetry, whereas no quantitative experiments in dynamics are described in Faraday's Chemical Manipulation(1827); see also T.H.Lodwig and W.A.Smeaton, 'The ice calorimeter of Lavoisier and Laplace', Annals of Science,XXXI(1974) 1-18.

13. H.Davy, Works,VIII,pp.172, 334ff.

14. H.Davy, Works, VIII,pp.287-8; a passage omitted in the quotation referred to Newtonian astronomy, and the hope that corresponding order would be traced in the terrestrial order of events.

15. The sterility of the Newtonian paradigm for chemistry by the time of Berthollet is urged by Thackray; Atoms and Powers (Cambridge,Mass.,1970) chap.7,3. Edinburgh Review,XI(1807) 390, XII(1808) 394; these and most of the other chemical articles were by Henry Brougham; Monthly Review LVII(1808)574. Reviews of Davy's papers and books are listed in J.Z. Fullmer, Sir Humphry Davy's Published Works(Cambridge, Mass.,1969); H.C.Oersted, The Soul in Nature,tr.L.and J.B.Horner (1852) p.301; H.Davy, Phil.Trans.,XCVIII (1808) 33; T.Beddoes, appendix I to J.E.Stock, Memoirs of the Life of Thomas Beddoes (Bristol,1811).

16. H.Davy, Phil.Trans.,C(1810) 16-37; on ammonium amalgam, Phil.Trans.,XCVIII(1808) 353-61; this substance was shown later to obey Boyle's Law, and is now generally regarded as a froth; C.Seely,

Chemical News,XXI(1870) 265. H.Davy,Phil.Trans.,
XCVIII(1808) 364 for the idea that all metals
contain hydrogen; cf Phil.Trans.,C(1810) 37. His
Bakerian Lecture of 1808 appears in Phil.Trans.,
XCIX(1809) 39-104.

17. H.Davy, Works,VIII,325; cf Phil.Trans.,C(1810)
69 — from Davy's Bakerian Lecture of 1809.

18. H.Davy, Phil.Trans.,XCVIII(1808) 368; Works,VIII,
p.323.

19. H.Davy, Works,II,p.23; Consolations (1830)pp.
195-6.

20. H.Davy, Phil.Trans.,X(1810) 239.

21. For reviews, see J.Z.Fullmer, Sir Humphry Davy's
Published Works (Cambridge,Mass.,1969); note The
Edinburgh Review,XVII(1811) 405, for some trepidat-
ion; Phil.Mag.,XL(1812) 145-51, 297-307, 434-44;
especially pp.439ff; for remarks on materialism, J.M.
Good, The Book of Nature (New York,1833) pp.48-9.

22. J.Davy, Fragmentary Remains ... of H.Davy (1858)
p.90; J.M.Good, The Book of Nature (New York,1833)
pp.102-3; H.Davy, Works,VIII, p.286; H.Davy, Elements
of Chemistry (1812) p.489.

23. J.von Liebig, Animal Chemistry, ed.W.Gregory
(Cambridge,Mass.,1842) p.195. On this general topic,
see J.H.Brooke, 'Wohler's Urea, and its vital force?
— a verdict from the chemists', Ambix XV(1968) 84-114.
Edmund Mills later gave a dynamical account of acidity
in which he claimed Davy as a predecessor: Phil.Mag.,
XXXVII(IVth series,1869) 461; 'the acidity of any
substance is a kind of resultant whose direction is
hydrogen'.

24. R.Cudworth, True Intellectual System (1678) pp.
29-38; J.Davy, Memoirs of H.Davy (1836) II,pp.95-6;
cf M.Faraday, The Non-Metallic Elements, ed.J.
Scoffern (1853) p.2; H.Davy, Consolations (1830); see
D.M.Knight, 'The Scientist as Sage', Studies in
Romanticism, VI(1967) 65-88; J.Davy, Memoirs of H.
Davy (1836) I,p.29.

25. C.Schreibers, 'Proteus Anguinus', Phil.Trans.,
XCI(1801) 241-64; E.Mendelsohn, Heat and Life
(Cambridge,Mass.,1964); and see D.M.Knight, 'The
Vital Flame',Ambix,XXIII,(1976)5-15. cf H.Davy,
Consolations (1830) pp.188, 206-7, 199-202, 211,
214-15.

26. S.Mauskopf, 'The Atomic Structural Theories of
Ampere and Gaudin:Molecular Speculation and
Avogadro's Hypothesis', Isis,LX(1969) 61-74; T.M.Cole,

'Early atomic speculations of M.A.Gaudin', Isis,LXVI
(1975) 334-60; B.C.Brodie, 'Ideal Chemistry',(1880)
p.13; the phrase does not appear in the contemporary
report in Chemical News,XV(1867) 296; W.H.Brock(ed.)
The Atomic Debates (Leicester,1967); see J.Z.Fullmer
Sir Humphry Davy's Published Works (Cambridge,Mass.,
1969) pp.98-100; W.Buchner(ed.), Goldkörner aus dem
literarischen Nachlasse eines christlichen Naturforschers (Erlangen,1856); B.J.Dendle tells us that
there were Spanish Romantics for whom temperament
and behaviour depended on electrical polarity:
Studies in Romanticism,VII(1968) 245; cf [R.Chambers]
Vestiges,5th edn(1846) p.348; see also T.H.Levere,
'Affinity or Structure:an early problem in organic
chemistry', Ambix XVII(1970) 111-26; and his Affinity
and Matter (Oxford,1971) chap.6; C.A.Russell, The
History of Valency (Leicester,1971).

27. J.P.Nichol, The Architecture of the Heavens,2nd
edn.(1850) pp.217, 219; the remark about tricks of
logic reminds us of the attack upon Laplace by W.
Whewell,Bridgewater Treatise,5th edn.(1836) p.333,
which provoked Charles Babbage's Ninth Bridgewater
Treatise (1837).

28. W.Prout, 'Gulstonian Lecture', London Medical
Gazette (1831) 257-65, 321-7, 385-91; see especially
p.261; and p.262 where he compares the atomic theory
to the Linnean System, artificial but convenient for
many purposes.

29. W.Kirby, Bridgewater Treatise,ed.T.R.Jones
(London,1853) I,pp.20-1, 33ff, on the cherubim as
powers; I,7,14 on Lamarck; I,p.347 on Jones, pp.20,
33 on Hutchinson; p.20 for Scriptural revelation.
See W.H.Brock,'The Selection of the Authors of the
Bridgewater Treatises', Notes and Records of the
Royal Society, XXI(1966) 162-179; C.C.Gillispie,
Genesis and Geology (Cambridge,Mass.,1951); and D.M.
Knight, Natural Science Books in English (1972)chap.3.
W.Kirby and W.Spence, An Introduction to Entomology
(4 vols,1815-26); this was used extensively by Darwin
— R.C.Stauffer, Charles Darwin's Natural Selection
(Cambridge,1975) 610.

30. W.Kirby, Bridgewater Treatise (1853) I,pp.21,
37-9, 40, 41, 52; II,p.176.

31. W.H.Wollaston, 'Bakerian Lecture',Phil.Trans.,
XCVI(1806) 13-22 on 'force' and 'energy' — see D.S.L.
Cardwell, 'Early Development of the Concepts of
Power, Work and Energy', British Journal for the
History of Science,III(1967) 209-24; J.Farey, A
Treatise on the Steam Engine (1827) pp.2, 16, 31ff;

this useful volume, and the previously unpublished
second volume, have been reprinted (Newton Abbot,
Devon,1971); W.E.Kirby, Bridgewater Treatise (1853)
II, pp.186-7; see D.M.Knight, 'Steps towards a
Dynamical Chemistry', Ambix,XIV(1967) 179-97; A.
Thackray, Matter and Powers (Cambridge,Mass.,1971);
R.E.Schofield, Mechanism and Materialism (Princeton,
N.J.,1970); J.E.McGuire, 'Force, Active Principles,
and Newton's Invisible Realm',Ambix XV(1968) 154-208;
and R.S.Fleming, 'Newton, gases, and Daltonian
Chemistry', Annals of Science,XXXI(1974) 561-74.

32. H.R.McAdoo, The Spirit of Anglicanism (1965)
chaps.3 and 4; H.More, The Complete Poems, ed.A.B.
Grosart (Blackburn,1878, reprinted Hildesheim 1969)
pp.55ff, 138; J.Glanvill, Saducismus Triumphatus,
3rd edn.(1689) p.224; R.Cudworth, True Intellectual
System (1678) pp.16, 11-13, 29-38.

33. W.Wollaston, Religion of Nature (1726) p.186;
J.Atkins, A Voyage to Guinea (1735) p.84; C.Maclaurin,
An Account of Sir Isaac Newton's Philosophical
Discoveries (1748) pp.388-9.

34. D.Hartley, Observations on Man (1749) I,p.364,
II,pp.31-4; J.Priestley, Hartley's Theory of the
Human Mind (1775) and Disquisitions (1777); R.Olsen,
'The Reception of Boscovich's ideas in Scotland',Isis,
LX(1969) 91-103; E.J.Mills,'The Atomic Theory', Phil.
Mag.,XL(IVth series,1871) 123; M.J.Nye, 'The nineteenth-
century atomic debates', Studies in the History and
Philosophy of Science,VII(1976) 245-68.

35. T.Reid, Essays on the Intellectual Powers of Man,
intr.B.Brodie,(Cambridge,Mass.,1969) pp.87, 109, 100,
xxxiv.

36. J.Davy, Memoirs of H.Davy (1836) I,p.37; on organs
as instruments, see H.Davy, Consolations(1830) p.210.

37. P.G.Tait and B.Stewart, The Unseen Universe (1876);
J.Tyndall, Fragments of Science, 10th imp.(1899) II,
pp.135-201; the address was delivered to the British
Association in Belfast in 1874. A.Gauld, Founders of
Psychical Research (1968); E.Gurney, F.W.H.Myers, and
F.Podmore, Phantasms of the Living (1886) xlix-lvi;
this work deserves to be better known; the spectacle
of hard-headed empiricists coming to terms with their
dubious and curious evidence is striking, and the list
of Members of the Council of the Society for Psychical
Research, pp.vii-ix, is impressive. There is an
entertaining review of P.G.Tait and Balfour Stewart's
Paradoxical Philosophy in The Scientific Papers of
James Clerk Maxwell,ed.W.D.Niven (Cambridge,1890)II,
pp.756-62. On natural theology in nineteenth century

Britain, see the unpublished Ph.D. thesis by Richard
Yeo (Sydney,1978)

4

The Fourth State of Matter

At the British Association Meeting in 1879 the bold
and wayward theorist William Crookes read a paper on
'Radiant Matter'.[1] When a potential difference was
applied across a gas at very low pressures, the
effects observed seemed qualitatively different from
what happened at moderate evacuation. As the pressure
fell, the dark spaces increased, the column of light
became striated and then disappeared, and a new kind
of radiation, causing phosphorescence when it struck
glass or crystals, became evident. Whereas at
moderately-low pressures, the discharge followed the
shortest path between the poles (and if deflected by
a magnet returned to its original path when beyond
the disturbance); the new kind of radiation was
propagated in straight lines from the negative pole,
and when deflectd by a magnet did not return to its
original path but continued in the new direction.
The famous experiments with a Maltese cross, a sharp
shadow of which was cast on the end of the tube, and
with paddle-wheels which turned when the radiation
fell upon their vanes, showed that the rays travelled
in straight lines, normal to the pole, and exerted
mechanical action. Crookes even demonstrated the
reaction on the pole; he was a brilliant exponent of
the lecture-demonstration, and one of the best-known
scientists of his day.

We cannot help interpreting these experiments in the light of J.J.Thomson's paper on cathode rays of 1897, in which he suggested that the rays were composed of corpuscles which might well be the prime matter.[2] Indeed, J.J.Thomson wrote:

> Thus on this view we have in the cathode rays matter in a new state, a state in which the subdivision of matter is carried very much further than in the ordinary gaseous state: a state in which all matter — that is, matter derived from different sources such as hydrogen, oxygen, &c. — is of one and the same kind; this matter being the substance from which all the chemical elements are built up.

But Helmholtz's Faraday Lecture on the particulate nature of electricity was not delivered until 1881, and there was no reason why Crookes should have identified his radiation with anything very like our electrons. He called it 'radiant matter', borrowing the term from one of Faraday's early lectures to the City Philosophical Society. Crookes argued that the radiation became apparent when the pressure was so low that the mean free path of the molecules was comparable with the dimensions of the tube; and, he added,' so distinct are these phenomena from anything which occurs in air or gas at ordinary tension, that we are led to assume that we are here brought face to face with matter in a fourth state or condition, a condition as far removed from the state of gas as a gas is from a liquid.' For Crookes, the passage of electricity illustrated this condition, but did not cause it.

In the ordinary way, according to Crookes, we have to think of gases in statistical terms; but at these low pressures we must:

> contemplate the molecules
> individually. In these highly-
> exhausted vessels the molecules
> dart across the tube with com-
> paratively few collisions, and
> radiating from the pole with
> enormous velocity they assume
> properties so novel and so
> characteristic as to entirely
> justify the application of the
> term borrowed from Faraday,
> that of Radiant Matter.[3]

Distinctions between different sorts of radiant
matter were hard to recognise, but Crookes thought
that molecules did retain their individuality. He
ended with a purple peroration:

> In studying this fourth state
> of matter we seem at length to
> have within our grasp an
> obedient to our control the
> little indivisible particles
> which with good warrant are
> supposed to constitute the
> physical basis of the Universe.
> We have seen that in some of
> its properties radiant matter
> is as material as this table,
> whilst in other properties it
> almost assumes the character of
> radiant energy. We have actually
> touched the borderland where
> matter and force seem to merge
> into one another, the shadowy
> realm between Known and Unknown
> which for me has always had
> peculiar temptations. I venture
> to think that the greatest
> scientific problems of the future
> will find their solution in this
> Borderland, and even beyond; here,
> it seems to me, lie Ultimate
> Realities, subtle, far-reaching,
> wonderful.

Crookes seems to have supposed that he was thus
supporting a brilliant insight of the young Faraday,

who had not himself been able to bring it to an experimental test. But in fact the terms 'radiant matter' and 'fourth state of matter' were by no means novel in Faraday's day; and light may be cast on the survival of Newtonian concepts if we look at the way in which these terms were used. We may then find it possible to distinguish physical arguments for the unity of matter from chemical ones, based upon relationships between atomic weights or between elements and radicals. Chemistry and physics to some extent went their separate ways during the nineteenth century, but such men as Davy, Faraday, and Crookes had a foot in both camps. For such people, the Newtonian chemistry remained attractive; involving imponderable fluids, a view of matter as composed of corpuscles in different arrangements, and the prospect of the quantification of short-range forces.[4]

Faraday, in his early lectures of which Crookes was so fond,[5] emphasised the unity of matter, in 1818 urging chemists 'to realise the once absurd notion of transmutation'. At the conclusion of a course of lectures, Faraday remarked that he had spoken of the 'powers and properties' of simple bodies rather than of the applications of chemistry; he had chosen to make his course 'elementary' and 'explanatory of the secret laws and forces on which the science, with all its uses, is founded'.

In the following year, Faraday discussed radiant matter, admitting it as hypothetical but justifying belief in it. The four states of matter — solid, liquid, gaseous, and radiant — differed in essential rather than secondary properties; and were 'most generally admitted. They do not belong to particular and separate sets of bodies, but are taken by most kinds of matter'. Some gases could not as yet be

liquefied except by adopting 'means which depend on
the rearrangement of particles'; that is, by chemical
combination. As we go from the solid through the
liquid to the gaseous condition, there is a progress-
ion; physical properties diminish in number and
variety. When solids melt, properties such as hardness and crystalline form disappear; gases 'form but
one set of substances, and the varieties of density,
hardness, opacity, elasticity and form, which render
the number of solids and fluids almost infinite, are
now supplied by a few slight variations in weight,
and some unimportant shades of colour'. Nature makes
'greater exertions' at each stage; and he felt that
it was not unreasonable to suppose that everything we
see may be converted into three kinds of radiant
matter. Philosophers failed to agree on the cause of
changes of state, and the status of heat and electricity; Faraday inclined to credit 'the immaterial nature
of these agencies'.

Crookes could have found all he needed in these
lectures of Faraday's;[6] but in Crookes' day talk of a
fourth state of matter must have seemed odder than in
Faraday's. The four states may take their origin from
the four ancient elements, but we shall not pursue
them further back than the late eighteenth century.
The critical figure in passing these ideas on to the
young Faraday was probably Davy; whose first published
paper does air the question of a fourth state, and of
the nature of heat and light.

This paper shows little sign of any influence
from Germany;[7] and indeed although Davy may be placed
among the Romantics, this does not imply that he was
seriously affected by Naturphilosophie, or by the
Kantian philosophy. The most influential prose work
of Coleridge was apparently his Aids to Reflection,

which consists of passages chiefly from the seventeenth-century Archbishop Leighton of Glasgow, with
commentary; and Coleridge is at least as important
for arousing interest in seventeenth-century divines
as for disseminating the doctrines of Kant. The
translations of Thomas Taylor made Alexandrian
Platonism available to English Romantics;[8] and of the
Cambridge Platonists Cudworth continued to be read,
his True Intellectual System being in the Royal
Institution's library by 1809. In reviewing Willich's
account of the critical philosophy, the Monthly
Review[9] compared the Kantians to the Neoplatonists,
charging both alike with obfuscation of the public
mind, and adding that the latter 'do not differ so
widely in spirit, as is commonly apprehended, from
those of the Konigsberg school'. Coleridge's annot-
ations to Taylor's translation of Proclus make the
same point; and Davy, in a lecture at the Royal
Institution in 1810, said that Ritter had gone astray
in following 'the metaphysical dogmas of Kant, which
as far as I can learn are pseudo-platonism', in place
of the doctrines of Bacon, Locke, and Hartley. Davy's
speculations about matter and spirit combined with
suspicion of systems are what one would expect in a
follower of Newton, or indeed of major trends in
seventeenth-century Anglican thinking. Coleridge
praised Bacon as the restorer of Platonic logic; and
Davy as the only exemplar of the Platonic-Baconian
method among contemporary chemists.[10]

It is difficult when confronted by someone like
Davy, who was not a philosopher, to separate Kantian
and other Romantic influences, even if one thinks this
a fruitful exercise. John Davy recorded[11] that his
brother had become acquainted with the Transcendental-
ists while still at Penzance; since he never learnt

German, there is some doubt as to the possibility
of this, but there were reviews of Kantian works in
the Monthly Review in the 1790s, one of which is
attributed to Beddoes, which show a limited sympathy.
Davy might have seen these through his first patron,
Davies Giddy, the former pupil and now friend and
correspondent of Beddoes who 'shared his literary and
chemical interests'. Davy, if he had read the
Monthly Review, would also have found there accounts
of the doctrines of the Scottish common-sense
philosophers, and of Mitchill's theory of contagion.

All this has diverted us from the paper of 1799
on the combinations of light. The combination in
this work of a corpuscular theory of light and a
kinetic theory of heat was far from novel; these
were typically Newtonian beliefs, and were so
described in Hutton's standard Mathematical and
Physical Dictionary.[12] Even the view of light as
capable of forming chemical compounds need not have
been startling; light had appeared in Lavoisier's
list of chemical elements, and members of the Lunar
Society had held ideas not unlike Davy's. Thus the
Monthly Review, commenting — favourably in the main
— on Davy's paper remarked that its basic tenet was
'by no means a new idea; but it is supported by the
author with great ingenuity, in the language of his
particular theory. ... We have seldom seen a fairer
promise of excellence than in these papers'. In the
same journal, a reviewer had commented on the essay
by one of the Gentlemen of Exeter that light seemed
to be 'a sort of favourite with this ingenious writer;
who scarcely sets any limits to its influence'.

Davy argued[13] that since light was not a modification of heat, and a continuous elastic aether an
impossibility, then light must be 'matter of a

peculiar kind, capable when moving through space with the greatest velocity, of becoming the source of a numerous class of our sensations'. Matter was possessed of the power of attraction, and there was also a repulsive power, that of heat. Heat 'may with propriety be called the repulsive motion'. These passages caught the attention of later important scientists, notably Thomas Graham the chemist and William Thomson (Lord Kelvin). Robert Angus Smith, the editor of Graham's works, wrote that Davy's ideas were 'a true movement in the direction of atomism', and that it had been the object of Graham's life, following Davy, to find what the movement of an atom was. Graham, he went on, was 'steadily thinking on the ultimate composition of bodies', and had concluded that while chemical atoms were indivisible by us, 'the original atom may be far down'. Rather differently, William Thomson in 1884 referred to Davy's 'suggestive statement' and 'most important idea'. Since he had read Davy's paper thirty-five years before, the idea had never left his mind that forces which were apparently-repulsive might in reality be the effect of attractive forces combined with inertia. The Joule-Thomson Effect — the cooling of gases by sudden expansion — was then the outcome of these attractive forces. Davy himself wrote that 'the particles of bodies may be considered as acted on by two opposing forces, the approximating power, which may (for greater ease of expression) be called attraction, and the repulsive motion', but his 'motion' seems to have been, like the Newtonians' and Graham's, vibrational whereas William Thomson's was translational — Davy's particles, that is, were running on the spot, and not dashing hither and thither, constantly colliding with each other.

Davy passed on to discuss the states of matter,[14] giving once again an account in line with the corpuscular philosophy. When the attractive predominates over the repulsive motion, then the substance is a solid; when they are in equilibrio, fluid; and when the repulsive motion is predominant, gaseous. But, he added, 'there is another state which has been heretofore unnoticed, and in which only one body that we are acquainted with exists, namely light'; and, he added in a footnote, probably odorous matters. In this fourth state, which can be called 'repulsive projection', the repulsive motion so predominates that the corpuscles seem to be very little acted on by attraction or gravitation. The differences in specific gravity between bodies are to be explained in terms of the different sums of the attractions and repulsive motions of their particles. The idea that matter and light are interconvertible is, as we noticed, to be found in the thirtieth query in Newton's Opticks; and with it the suggestion that bodies owe much of their activity to combined particles of light.

According to Lavoisier[15] gases were compounds of caloric; and the antiphlogistic chemists therefore spoke of 'oxygen gas' and of its hitherto-unknown basis; combustion was the decomposition of oxygen gas. To Davy this was improper; hydrogen and nitrogen were simple substances 'relative to the present state of our knowledge', while (since combustions were characterised by the emission of light) oxygen gas was a compound of light. Having disposed of caloric, Davy sought to show that there was no analogy between changes of state and chemical reactions. Bodies were equally simple in each state; simple substances should be given names which did not include a

99

description of their state at common temperatures. In a footnote he remarked that hydrogen and nitrogen were probably metals in a state of vapour; to this speculation he later returned, as we have seen in the discussions of ammonium amalgam; and in 1844 Faraday declared that 'the hope of seeing nitrogen, oxygen, and hydrogen, either as liquid or solid bodies, and the latter probably as a metal, has lately induced me to make many experiments on the subject'.

It has been suggested,[16] both at the time and more recently, that Davy's researches can best be understood as a reaction to the system of Lavoisier; as an attempt, that is, to prove that chemistry was not a French science. Dumas, who argued thus, added that Davy had been carried away by alchemical and pantheistic ideas. While the Royal Institution acquired in Davy's day a number of alchemical books, and both Davy and Faraday expressed the view that the alchemist had been right in principle, Davy's series of researches fit not unhappily into a Newtonian frame-work. His first paper showed that heat was motion; his researches on oxides of nitrogen showed what a range of properties compounds of only two elements could display; his investigations in electrochemistry promised an understanding — and eventually a quantifi-cation — of short-range forces; his work on chlorine showed that acidity did not depend upon the presence of property-bearing principle; and his studies of the diamond and charcoal proved that substances chemically identical could display very different physical properties, which must therefore be the consequence of arrangements of particles. In believing that heat and light were probably ordinary matter, and in being apparently uncertain whether the active powers were aethereal forms of matter or immediate expressions of

100

the Divine Will, Davy was again close to the Newtonians. While he certainly retracted some of his
remarks of 1799 along with his new nomenclature, and
expressed a resolve to reject hypotheses in the
future; it is not necessary to take this too seriously,
or to suppose that he was shortly afterwards awakened
from his Hartleyan dogmatic slumbers by specifically
Kantian philosophy.

Davy is particularly hard to follow because of
his tendency to see all sides of a question, his lack
of interest in consistency, his opposition to systems,
and his originality. But the idea that light was a
form of matter, 'radiant matter', is to be found as a
received doctrine in standard textbooks of the first
decades of the nineteenth century, along with a
corpuscularian account of changes of state and of
chemical combination in terms of the 'attractions' of
aggregation or cohesion, and of affinity. Most of
these textbooks differed from Davy in taking for
granted the caloric theory of heat; and indeed after
William Herschel's discovery of infra-red radiation
it became impossible to regard radiant heat and light
as very different from one another.[17] If heat were a
vibration, so must light be; or both must be minute
particles. Herschel's own first explanation was that
radiant heat consisted of particles of light having
a different range of momenta from those constituting
visible light. He was careful to distinguish the
sensation, heat, from its cause, for which he believed
that a different word was required, the usual
candidate, caloric, being unacceptable. The rays
which occasion heat are not themselves heat. This
separation of cause and effect was traditional.

Parkes, in successive editions of the elementary
Chemical Catechism[18] between 1806 and 1822 had no

hesitation in describing caloric as the cause of the
sensation of heat. He distinguished 'sensible' from
'latent' heat, using the analogy of damp bread, from
which some of the water can be squeezed, while some
is chemically combined in the carbohydrate. Caloric
is similarly retained in bodies by affinity. To the
famous Mrs Marcet, in her informal Conversations on
chemistry, caloric was a fluid composed of minute
particles constantly in agitation; the two principle
solvent fluids in chemistry were water and caloric.
A 'gas' was distinguished from a 'vapour' in which
the caloric was less firmly combined. In nature we
see 'a kind of continual warfare between the attraction of aggregation and the repulsive power of caloric'.
Mrs Marcet's atoms were the particles of the
corpuscularians, which composed all matter so that
'a pound of feathers does not contain one atom more
than a pound of lead'.

In Thomas Thomson's standard and more-formal
textbook A System of Chemistry,[19] we find a possible
origin for Faraday's idea that there might be three
different kinds of radiant matter. In the edition
of 1804 he wrote that Herschel's experiments had
strengthened the view that heat was material; and
that its particles must be as minute as those of
light, and hence appear weightless to us. He added:
'Solar light is composed of three distinct substances,
in some measure separable by the prism on account of
the difference of their refrangibility. The calorific
rays are the least refrangible, the deoxidizing rays
are most refrangible, and the calorific rays possess
a mean degree of refrangibility'. The ultra-violet,
or deoxidizing rays had recently been discovered by
Ritter. Thomson's separation into three kinds of
rays was more thoroughgoing than that of many of his

contemporaries; for there was much discussion around
the year 1800 of how many primary colours, and therefore perhaps different species of light, there were.
Wollaston's paper in which he described some of the
lines in the solar spectrum to which Fraunhofer later
gave his name, was an attempt to elucidate this
question; and so was an interesting book by James
Sowerby, first of a line of very distinguished
illustrators of natural history, who was chiefly concerned with drawing up a natural classification of
colours, and a system of standards, based upon three
primaries.

Thomson remarked that the study of these 'bodies'
had so far made little progress; but added that 'we
may look forward to this subject as likely to correct
many vague and unmeaning opinions which are at
present in vogue among chemists'.[20] He believed that
even the conduction of heat was an effect of affinity,
a succession of equilibria being established through
the body. In William Thomas Brande's compilation, A
Manual of Chemistry, of 1819 we find the heading 'Of
Radiant Matter', introducing a discussion of Newton's
theory of particles, and Descartes' and Huygens' of
an elastic medium. Brande, who was Davy's successor
at the Royal Institution, thought that the phenomena
were best explained in terms of matter in motion, and
raised Newton's Query 30, on whether light and matter
were interconvertible, with a reference to Davy.
Like Thomson, he wrote that the solar beams were
refrangible into three distinct kinds, calorific,
luminous, and decomposing rays. And J.B.Daniell —
a chemist famous for his electro-chemical work — in
1839 still presented the undulatory and emission
theories of light as alternatives; remarking that
according to the latter, particles of 'prodigious

self-repulsive force' were projected from luminous
bodies at an inconceivable velocity. His Irish contemporary Sir Robert Kane in his textbook, which was
very popular in the USA, wrote of the balance between
affinity and the repulsion due to heat, which led to
reversibility and equilibria in chemical reactions.
It is surprising in view of these survivals to find
that quantitative studies of heat changes in chemical
reactions do not seem to have been widely attempted in
England in the first half of the nineteenth century;
one of the major differences between Lavoisier's and
Faraday's writings on chemical manipulation is that
the latter offer no advice on calorimetry. One is
tempted to suppose that chemists skipped the theoretical introductions to their textbooks and went straight
to the detailed and practical parts.

Another textbook — less important — which gives
relatively wide coverage to the 'ethereal chemical
elements' (heat, light, electricity, and magnetism)
was written by Weldon and published it seems in 1824.[21]
Some philosophers, he wrote, considered heat and light
as being properties arising out of unknown modifications of matter; others as being an active aethereal
kind of matter; and others again 'as the common material
elements in a peculiar state; which may be contrasted
to solid matter in its usual state as the state of gas
is'. This is not exactly clear, but seems to be a
reference to a fourth state of matter of great tenuity.
Heat and light, he wrote, appear to enter into chemical
combinations in fixed quantities:

> But their inconfinable nature;
> their want of gravity; their un-
> accountable transference, or
> accumulation upon various occasions;
> the singular affections they exert;
> and the protean states they

communicate to matter; their
mutual influence too over each
other; their analogy to some of
the chemical elements, and to
each other in some properties,
do not allow any conclusion as
to their nature. It appears
more logical in the present
state of our knowledge, to consider them as distinct elements.
Whether they are called material
or not is of little consequence,
since we have still to learn what
matter is. It is by no means
necessary that every element that
exists, and is not mental, should
consist of impenetrable extended
gravitating atoms ... it is very
possible that various elements or
agents were created; differing in
their properties ... and that
these agents or elements, by the
various modifications arising from
their respective actions on each
other, produce the bodies and
phenomena around us.

For material substances, gravity would then be an inherent and necessary property; immaterial substances
would be devoid of gravity and impenetrability; and
heat, light, magnetism, and electricity would form a
chain linking the two classes.

The physician John Mason Good lectured at the
Surrey Institution in the second decade of the nineteenth century, and his lectures were later published
as The Book of Nature.[22] We learn something of his
outlook on life from the biographical sketch in the
posthumous American edition of this book. Thus Newton's
Principia and Opticks had for him an authority only
just below that of the Bible; and when in 1807 the
Minister of the Unitarian church Good was in the habit
of attending declared in a sermon that the traditional
arguments did not prove beyond doubt the existence of
God, he wrote stiffly to him that he had learnt a

different creed 'from Sir I.Newton, Clarke, Barrow, and Locke', and that he could not 'without pain, hear what appears to me a principle irrefragibly establish- ed, treated with skepticism', and joined a Trinitarian congregation. He was avowedly anti-Platonic, and took a high view of matter as bearing 'indubitable proofs of being capacified for incorruptibility'. The theory of Boscovich, which he says in a lecture was generally adopted, was 'fine-spun and fanciful'; we must take as our basis 'solid, impenetrable, and moveable particles'. For Good, heat, light, electricity, and magnetism possessed none of the common properties of matter; they were powers which we could only know by their effects, and their particles repelled each other. Matter in its simplest state was possessed of certain active, universal, and essential properties, such as repulsion and gravitation; but whether they were inherent or not was an insuperable problem. Common matter had, in peculiar states of modification, 'peculiar properties, independently of the general or essential properties which belong to the entire mass', such as magnetic or electric charge. The soul probably consisted of 'spiritualized, ethereal, gaseous or radiant matter'.

As well as believing that souls were composed of radiant matter, Good expressed great interest in William Herschel's hypothesis that light was the basis of ordinary matter.[23] This is to be found in various speculative papers in the Philosophical Transactions between 1801 and 1814, addressed to 'any astronomer, chemist, or philosopher' in the belief that the subjec 'cannot fail to attract the notice of every inquisitiv mind'. Herschel's paper of 1801 was an expansion of his suggestion that the Sun was inhabited, the sun- spots being openings through which the cool interior

106

could be glimpsed. Such openings would be impossible
if the Sun were liquid — as Hygens had urged — or
gaseous; and it was impossible, despite the 'seeming
impropriety', to 'refrain from ideas that would
obtrude themselves', such as that the 'shining matter'
of the Sun must 'exist in the manner of empyreal,
luminous, or phosphoric clouds, residing in the higher
regions of the solar atmosphere'. Further papers in
1811 and 1814 presented the nebular hypothesis: that
diffused shining matter was condensed by gravitational forces into globular forms, the collapse of branches
producing vortical motions; and that the final result
of the process was the production of stars and
planetary systems. In 1814 Herschel published
another paper to illustrate the 'ultimate connection
between the nebulous and sidereal condition' through
a series of ambiguous objects.

Norman Lockyer's spectroscopic observations on
sunspots established that they were not openings
through a fiery atmosphere to a cool interior; and the
resolution of various nebulous bodies by Lord Rosse
with the great reflector at Parsonstown in Ireland
brought into disrepute the nebular hypothesis of the
origin of stars.[24] When in the 1860s and 1870s
hypotheses of inorganic evolution were revived —
following the work of Bunsen and Kirchhoff, of Miller
and Huggins, and of Lockyer, with the spectroscope —
the connection with the emission theory of light was
no longer there. The American geochemist Sterry Hunt[25]
did however propose in 1882 that the aether of Newton's
second paper on light and colours might be the food of
the planets and the material basis of life; quoting
from Newton the suggestion that 'the whole frame of
nature may be nothing but various contextures of some
certain etherial spirits or vapours condensed ... much

after the manner that vapours are condensed into water'. This is perhaps not far from the view that the aether is a fourth state; and the lecture was printed in Crookes' journal Chemical News. The idea that the aether might be constituent of ordinary matter was presented to the British Association in 1886 by Professor Carnelly of Dundee; his aether had an atomic weight of -2, and with carbon and hydrogen composed all substances. This speculation at this late date is curious, but to dignify it with the epithet 'Newtonian' would be absurd.

Rather more important were the views of the physician William Prout,[26] whose name is associated with the view that all the elements are as it were polymers of hydrogen. His early speculations — in student essays at Edinburgh — show the typical Newtonian preoccupations of the opening years of the nineteenth century; his famous Hypothesis grew out of this and the 'volumes and atoms' controversy which followed the publication of Gay-Lussac's Law; and his Bridgewater Treatise on natural theology — at least the chemical part of it — represents an attempt to provide an explanation of polarity, which Davy had shown to be the basis of chemical affinity, in terms of corpuscles or molecules. He remarked that the three states of matter shade into one another, 'vapour', for example, referring to an imperfect gas; and under the heading of gases he included vapours and 'the etheriform condition of matter'. It had been suggested that molecules had the same shape as the crystals composed of them; but this seemed 'so unlike the usual simplicity of nature's operations' that it was more plausible to suggest that the molecules were spheroids. Prout's view was typical of English crystallographers, while the alternative he rejected was popular in France. Polarity, a molecular rather than universal agency, had then to be invoked to explain cohesion; electricity probably corresponding to chemical, and magnetism to cohesive polarity.

In liquids the particles were loosened by heat; but 'heat' is the name of a sensation, the cause of which was in doubt. Prout believed[27] that heat and light were 'substances, the molecules of which are influenced by polarizing forces, precisely similar in all respects, to those which influence common matter'. In a footnote he asserted that the chemical action of light was to be explained on chemical principles, whatever mathematicians might suppose about light being undulations; chemists and physicists, as we have noted, often viewed one another with suspicion at this period. The heat absorbed by bodies formed, in Prout's view, an atmosphere around their particles; latent heat being converted into polarizing energies. The axes of polarity in the molecules of solids are parallel; in liquids, at right angles; and in gases, parallel but opposed so that the molecules repel each other as in experiments on diffusion. From this theory of gases, Prout derived 'Avogadro's Hypothesis', that equal volumes of all gases contain equal numbers of molecules under the same conditions; assuming that gases are composed of molecules — which they must be to exhibit polarity — obey the gas laws, and combine according to Gay-Lussac's Law.

From the high velocity of light, Prout deduced that it must, if matter, be in an exceedingly tenuous form; probably 'polarized molecules in the self-repulsive state', which might therefore obey the gas laws.[28] Radiation is thus analogous to gaseous diffusion, and comparison of the rates of the two

109

processes might be used to estimate the density of light. The axes of polarity required to explain chemical affinity could when applied to light account for the Newtonian fits of easy transmission and reflection. Following Davy, Prout suggested that the differences between carbon and the diamond illustrated how properties depend upon the arrangements of particles rather than on their nature. The homologous series of organic Chemistry showed how the number of units in a compound affect its properties; and Prout concluded that the

> sub-molecules may be imagined to
> resemble in some degree, the
> imponderable matters, heat &c.,
> not only in their extreme tenuity,
> but in other characters also; and
> this very intensity of property
> and character may be reasonably
> considered as one, if not the
> principle reason, why they are in-
> capable of existing in a detached
> form. Lastly, are not these
> ultimate and refined forms of
> matter extensively employed in
> many of the operations of nature;
> and particularly in many of the
> processes of organisation?

The real elements of bodies, the sub-molecules, are thus very like heat and light; but they do not seem to be the same.

The terms 'radiant matter' and 'fourth state of matter' which we found in Faraday and in Crookes, and which imply that matter and light are interconvertible, are however to be found in the prospectus seeking subscriptions for a book on 'The Elements of Radiant and Fixed Matter' which was described at some length in the Philosophical Magazine for 1819.[29] The view presented was that matter existed in four forms, the first three of which (our three states) were passive:

changes in the universe were to be ascribed to
radiant matter, aided by caloric. Light, for this
would-be author, was a 'material compound, composed
of the four simple elementary principles, or un-
decompound constituents of matter, of which all other
bodies in nature are formed'. By a prism light could
be divided into four 'original prismatic coloured
rays' which by obliquity of position generate all
other colours. These primary rays 'possess peculiar-
ly distinct and countervailing qualities, and on the
proportions in which they are combined in matter, and
the nature of the polarity exercised in their combination, its specific properties are totally dependent'.
The red ray has acidifying and oxygenating powers;
the yellow was nitrogenous and alkalescent; the blue
was the carbonic ray, and the blue hydrogenating.
They combined electively, according to laws, with
inert matter. At low temperatures, caloric assisted
the combination, but above 800-900°F it effected the
restoration of matter to the radiant state. The light
particles were spherical — the young Faraday adhered
to the more up-to-date view of Malus, for whom they
were octohedral — being quaternary compounds with one
positive and three negative poles at the centre, and
one negative and three negative poles at the surface.
The galvanic phenomena indicated too that the transition to the radiant state was effected when the
positive pole of an oxygen atom came into contact with
negative poles of carbon, nitrogen, and hydrogen atoms,
and combination occurred; the process being reversed
on their combination by opposite poles. It is perhaps
not surprising that nothing seems to have come from
this prospectus.

All these examples seem to indicate that the
belief that matter could be brought into a radiant or

fourth state was a part of the Newtonian or
Corpuscularian legacy to which the natural philosopher
or chemist was heir. This inheritance is obvious in
the introductory passages of the textbooks of the
early nineteenth century, with their discussions of
the 'attractions' of aggregation and affinity. Only
after the passage of two or three decades of the
century, when the study of galvanism had failed to
bring about a quantification of short-range forces,
the emission theory of light had given way to the
undulatory, and chemists had become increasingly
involved in the organic branch of the subject, did
the Newtonian tradition wither away; so that to
Crookes with his cathode-ray tube Faraday's early
lectures seemed a vision of the future when in fact
they were a survival from the eighteenth century.
The corpuscular philosophy was remote from a world of
aethereal vortices and chemical atoms as it had not
been in the days of Davy and Berthollet.

We have seen how Davy's researches can be
described in Newtonian terms, properties depending
'upon different arrangements of the same particles of
matter' under the direction of active powers;[30] this
may be contrasted with Priestley's belief in active
matter, and with the Kantians' belief in the reality
only of force. A great problem on which these
doctrines were expected to cast light was that of
chemical synthesis. John Stuart Mill distinguished
in his Autobiography between mechanics and chemistry;
in the former science forces could be summed whereas
in the latter they could not: 'a science is either
deductive or experimental, according as, in the
province it deals with, the effects of causes when
conjoined, are or are not the sums of the effects
which the same causes produce when separated'. Mill

112

adds that he developed this distinction from a remark 'in the introduction to that favourite of my boyhood, Thomson's System of Chemistry'. The properties of water are clearly not the sum or average of those of oxygen and hydrogen; a genuine synthesis, with the formation of a tertium quid has taken place. This had been stressed by Coleridge, following Schelling; and Mill's contemporary William Whewell, who like Coleridge was au fait with some German philosophy, was to dilate upon the folly of trying to reduce chemistry to mechanics since both sciences were erected upon different fundamental ideas. But a doctrine which Mill derived from Thomas Thomson cannot with great plausibility be given a German pedigree.

Dalton's theory seemed unable to explain chemical synthesis, which required more than juxtaposition of particles as induction requires more than juxtaposition of facts; but there were few in the nineteenth century who opted for the Plotinian doctrines of the continuum and complete interpenetration. Corpuscularian opposition, on the other hand, to the homoiomeria of Anaxagoras provided ammunition for those who disliked the idea that the chemical elements might be composed of atoms possessing the chemical properties of the elements themselves — which could be seen as an up-to-date version of Anaxagoras. Such a theory could not account for chemical synthesis; and indeed if it be the object of an atomic theory to explain the properties of things in terms of the arrangements of atoms having simpler properties, then Dalton's, however useful in other ways, was not really an atomic theory at all. Thus in a dialogue of Davy's we find: 'You mistake me if you suppose I have adopted a system like the Homooia (sic) of Anaxagoras, and that I

suppose the elements to be physical molecules endowed
with the properties of the bodies we believe to be
indecomposeable'.[31] To the corpuscularian, matter
was a veritable Proteus; and the powers that modify
matter were the proper field for his researches.
This did not necessitate the view that nature was an
incomprehensible flux, for Newton had shown the
simplicity and harmony of nature. He had brought
order into mechanics and astronomy; and it was therefore reasonable to hope with Davy that the laws of
affinity might be subjected to mathematics and 'the
possible results of new arrangements of matter become
the objects of calculation'. There was even hope of
one great general law which would unify all the
phenomena; but here again we do not need to invoke
Naturphilosophie when we have Maclaurin to tell us
that Newton was induced, following a maxim ascribed
to the legendary founder of alchemy, Hermes
Trismegistus, 'to think that all these powers might
proceed from one general instrument or agent, as
various branches from one great stem'.

With this warning against underrating the
corpuscular philosophy in Regency England, we can
glance at the work on the discharge tube by Crookes'
predecessors, Davy and Faraday. In 1822 Davy
published a paper 'on the electrical phenomena
exhibited in vacuo'. It begins:

> The production of heat and light
> by electrical discharge; the manner
> in which chemical attractions are
> produced, destroyed, or modified
> by changes in the electrical state
> of bodies; and the late important
> discovery of the connection of
> magnetism with electricity, have
> opened an extensive field of
> inquiry in physical science, and
> have rendered investigations,

114

concerning the nature of
electricity, and the laws by which
it is governed, and the properties
that it communicated to bodies,
much more interesting than at any
former period in philosophy. Is
electricity a subtile elastic
fluid? — or are electrical effects
merely the exhibition of the
attractive powers of the particles
of bodies? Are heat and light
elements of electricity, or merely
the effects of its action? Is
magnetism identical with electricity,
or an independent agent, put into
motion or activity by electricity?[32]

Such questions 'few sagacious reasoners' would wish
to answer in the present state of knowledge 'on such
very abstruse and difficult parts of corpuscular
philosophy'. Davy hoped that studies of the relation
of electricity to void space might be helpful; his
discharge tube had a vacuum over cooled mercury and
other metals, and he observed phosphorescent 'electric
light'. He thought that while the light depended
'principally upon some properties or substances
belonging to the ponderable matter through which it
passes', space was capable of exhibiting electrical
phenomena. This was compatible with electricity being
either a particulate fluid or an aetherial matter as
proposed by Hooke, Huygens, and Euler; or with the
view that positive and negative electrical states may
increase the force of vapour from bodies, which would
explain why the poles of the carbon arc may be drawn
further apart when once it was set going. This last
view does not commit one to any theory of the nature
of electricity.

These inconclusive questionings are reminiscent
of those of the early years of the century when Volta's
discovery opened new and strange vistas. What does

seem clear is that Davy was not concerned with whether electricity could bring matter under low pressure into a fourth state, but with whether electricity was matter. We may compare Davy's paper with one of 1785 by William Morgan,[33] using similar apparatus. Doubtless to his relief, Morgan found that a vacuum did not conduct; for if it had, then either 'our atmosphere must have been long ago denuded of all its electric fluid by being everywhere surrounded by a boundless conductor', or else there would be no vacuum in the universe — a terrible thought to an eighteenth-century Newtonian. Morgan found that even imperfect vacua did not conduct, and concluded 'that there is a limit even in the rarefaction of air, which sets bounds to its conducting power; or in other words, that the particles of air may be so far separated from each other as no longer to be able to transmit the electric fluid'. There might be another limit when the air was so compressed 'as to resist the passage of the fluid entirely'.

Davy was less interested in atmospheric electricity, and unhappy with the doctrine that electricity was a fluid; and so was Faraday,[34] whose experiments on electricity in vacuo come in the thirteenth series of his Experimental Researches. He described the positive column, the negative glow, and the dark space named after him; but again his interest was in the nature of electricity and of action apparently at a distance, and not on the state of the matter in the tube — although the experiments follow others on electrical conduction by streams of dust particles. Faraday remarked that it would be interesting to know whether a perfect vacuum would be a conductor or a dielectric, or something different from both; and referred to the papers of Davy and of Morgan.

So it seems that we can assign to Crookes the credit for applying for the first time in the 1870s what would twenty years before have seemed an exploded theory of the nature of light, to the phenomena of the discharge tube; and for giving an explanation of the effects which, although it did not survive, paved the way for the 'corpuscular' account of J.J.Thomson which prevailed in place of the widely-favoured aetherial explanations.[35] To say this is not to aver that Crookes knew the history behind the concept of a fourth state of matter; nor is it to deny that there were features of his account derived from the kinetic theory of gases which nobody could have used in the early nineteenth century. But the story may serve to show some links between the corpuscles of Newton and Boyle, and those of J.J.Thomson; different aggregations of which formed the various chemical elements. We now pass on to see how the belief in the unity of matter received support from chemical analogies.

NOTES

1. W.Crookes, 'On Radiant Matter', Nature,XX(1879) 419-23, 436-40; R. de Kosky, 'William Crookes and the fourth state of matter', Isis,LXVII(1976) 36-60.

2. J.J.Thomson, 'Cathode Rays', Phil.Mag.,XLIV (5th series,1897) 293-316, 312; H.von Helmholtz, 'Faraday Lecture', Nature,XXIII(1881) 537-40 — this is an abbreviated form of the lecture, which was published in full by the Chemical Society; H.Bence Jones, Life of Faraday, 2nd edn.(1870) pp.269-71; Crookes, Nature,XX (1879) 419.

3. W.Crookes, Nature,XX(1879) 420, 439-40. Crookes pursued these studies, and described them before the Institution of Electrical Engineers, of which he was President, in 1891: Chemical News,LXIII(1891) 53-6, 68-70, 77-80, 89-93, 98-100, 112-14. Crookes compared the cathode rays to a stream of bullets from a machine gun, Nature,XX(1879) 437; cf the experiments of J.Mitchell on light corpuscles reported on by J.Priestley, History of Vision, Light and Colours(1772)

pp.387-90. On the kinetic theory of gases, see E.
Mendoza, 'Herapath's theory of gases', British
Journal for the History of Science,VIII(1975) 155-65;
S.G.Brush, Kinetic Theory (Oxford,1965-6); G.R.Talbot
and A.J.Pacey, 'Some Early Kinetic Theories of Gases:
Herapath and his predecessors', British Journal for
the History of Science,III(1966) 133-49; E.E.Daub,
'Waterston, Rankine, and Clausius on the Kinetic
Theory of Gases',Isis,LXI(1970) 105-6; E.W.Garber,
'Clausius and Maxwell's kinetic theory of gases',
Historical Studies in the Physical Sciences,II(1970)
299-30; C.C.Gillespie, 'on analysis by probabilities',
in A.C.Crombie(ed.),Scientific Change(1963) pp.431-53.

4. T.H.Levere, Affinity and Matter(Oxford,1971)chap.
7; see also R.Fox, 'The Rise and Fall of Laplacean
physics', Historical Studies in the Physical Sciences,
IV(1972) 89-136;"The Laplacian Programme for Physics",
Bolotin de la Academia Nac. de Ciencias,Cordoba,
Argentina,XLVIII(1970) 429-37.

5. H.Bence Jones, Life of Faraday,2nd edn.(1870) I,
pp.225-7, 269-73; T.H.Levere, 'Faraday, Matter, and
Natural Theology — reflections on an unpublished
manuscript', British Journal for the History of
Science,IV(1968) 95-107; P.M.Heimann,'Faraday's
Theories of Matter and Electricity', British Journal
for the History of Science,V(1971) 235-57; L.P.
Williams, Michael Faraday(1965) chaps.1 and 2; and see
W.V.Farrar's article on Faraday and Richard Laming,
Annals of Science,XXV(1969) 243-54.

6. Crookes quoted Faraday's early lectures both in
1879 — Nature,XX(1879) 419, and in his Presidential
Address to the Chemical Section of the British
Association; B.A.Report,(1886) 559. Faraday's remarks
about the greater exertions made at each stage
presumably means that the latent heat of vaporization
is greater than that of fusion. See J.L.Gay-Lussac's
paper 'On the Simplicity of the Gaseous State', in
Alembic Club Reprints,IV(1961) 8-24, showing how gases
combine in simple ratios by volume. See also T.G.
Collingwood, The Idea of Nature(Oxford,1945) p.78 on
the four elements; p.415 on the gulf between physics
and chemistry. H.Davy, Works,II,pp.5-86; VIII,pp.335-
6 on radiant matter.

7. See D.M.Knight, 'Physical Sciences and the
Romantic Movement', History of Science,IX(1971) 54-75;
T.H.Levere, Affinity and Matter(Oxford),1971) chap.2,
presents a more-Kantian Davy. S.T.Coleridge, Aids to
Reflection,ed.T.Fenby (Liverpool,1874) is the edition
I have used; a critical edition will appear in the
Collected Works. D.Newsome, Godliness and Good

Learning(1961) pp.14ff, 196ff.

8. K.Raine and G.M.Harper, Thomas Taylor the
Platonist:Selected Writings, (1969); W.Enfield, whose
History of Philosophy from the Earliest Times to the
Beginning of the Present Century(1791) we know Davy
read, described the Alexandrian Platonists with
fascinated horror.

9. Monthly Review,XXVIII(1799) 65; S.T.Coleridge,
Notebooks,ed.K.Coburn,(1957) I,pt.2,p.457; H.Davy,
Works,VIII,p.272; H.R.McAdoo, The Spirit of Angli-
canism(1965)ch.2 for opposition to systems; chaps.7
and 8 on the new philosophy and theological method.

10. S.T.Coleridge, Notebooks,ed.K.Coburn(1957) I,pt.2,
p.459, and The Friend, ed.B.S.Rooke(1969) I,p.482,471;
see also the passages on pp.422 and 530, on French
materialism in science and on the idea that Davy was
successful in view only of his splendid apparatus and
his industry; S.T.Coleridge, Hints Towards the
Formation of a more comprehensive Theory of Life,
ed.S.B.Watson(1848) p.23; cf Georges Cuvier, 'Eloge
de Davy', Memoires de l'Academie Royale des Sciences,
XII(1833)xxxv.

11. J.Davy, Memoirs of H.Davy(1836) I,pp.37-8; S.T.
Coleridge, Notebooks, ed.K.Coburn,I,sections 249 note
and 1326 note: T.H.Levere, Affinity and Matter
(Oxford,1971)p.29.

12. C.Hutton, A Mathematical and Philosophical
Dictionary(1795-6) II,pp.33, 35 on light; I,pp.587-8
on heat; James Watt's letter to Joseph Black, S.
Smiles, Lives of Boulton and Watt(1865)p.380; Monthly
Review,XXX(1799)60-72 — see especially p.65; Monthly
Review,XXII(1797)1-10. Other reviews were less
indulgent, but the reception of the paper does not
seem to have been generally hostile. See J.Z.Fullmer,
Sir Humphry Davy's Published Works(Cambridge,Mass.,
1969)27-8, and H.Hartley, Humphry Davy(1969)p.19.

13. H.Davy, Works,II,pp.8-9, 10-11, 14-15; cf C.
Hutton, Mathematical and Philosophical Dictionary
(1795-6) II,p.336; T.Graham, Chemical and Physical
Researches,ed.R.A.Smith (Edinburgh,1876)pp.xvi-xix;
W.Thomson, British Association Report,(1884)614; this
was a Presidential Address to Section A of the
Association. The papers of Joule and Thomson 'On the
Thermal Effects of Fluids in Motion' are conveniently
accessible in J.P.Joule, Scientific Papers,II(1887)
pp.231-362; Joule elsewhere (Scientific Papers,I(1884)
p.266, suggests however that repulsion may be a
faculty with which matter is endowed. See Graham's
'Speculative Ideas respecting the Constitution of

Matter', Phil.Mag.,XXVII(4th series 1864)81-4. See
also on the antecedents of kinetic theory W.L.Scott,
The Conflict between Atomism and Conservation Theory
(1970), especially on Joule; H.I.Sharlin, 'William
Thomson's dynamical theory', Annals of Science,XXXII
(1975)133-48.

14. H.Davy, Works,II,p.16; cf C.Hutton, Mathematical
and Philosophical Dictionary(1795-6)II,p.336; I.
Newton, Opticks, 4th edn, reprinted (New York,1952)
p.374. H.Davy, Works,VIII,p.341-2 discusses whether
heat and light are ordinary matter, or some specific
kinds.

15. A.L.Lavoisier, Elements of Chemistry(1790)p.78;
H.Davy, Works,II,pp.21-2; M.Faraday, Experimental
Researches in Physics and Chemistry(1859)p.96.

16. See R.Siegfried's introduction to the reprint
of Davy's Works(New York,1974); J.B.Dumas, Lecons sur
la Philosophie Chimique,ed.M.Bineau(Paris,1837)IXme
leçon; H.Davy, Works,VIII,pp.328-31; M.Faraday,
Lectures on the Non-metallic Elements(1853)p.7; for
remarks on 'pwers' in Davy's lectures, see his Works,
VIII,pp.168-72, 287-8, 336-7, 345-51; the last of
these contains a warning against notions, doctrines,
and dogmas like that against hypotheses in C.Hutton,
Mathematical and Philosophical Dictionary(1795-6) I,
p.621; Davy's recantation is in Nicholson's Journal,
III(4th series,1800)517. On the survival of
Renaissance ideas in Chemistry, see A.Debus, Chemical
Philosophy(New York, in press).

17. S.G.Brush, 'The Wave Theory of Heat', British
Journal for the History of Science,V(1970)145-67;
D.J.Lovell, 'Herschel's dilemma in the Interpretation
of Thermal Radiation',Isis,LIX(1968)46-60; Herschel's
experiments appeared in Phil.Trans.XC(1800); I have
used the Abridgement of the Phil.Trans.,ed.C.Hutton
et al.,(19 vols,1809)XVIII,pp.683, 692. In C.Hutton's
Mathematical and Philosophical Dictionary(1795-6) I,
687, we find the same distinction between sensations
and their causes.

18. S.Parkes, Chemical Catechism(1806)p.103; J.Marcet,
Conversations on Chemstry(1806) I,pp.21-2, 25, 62; the
experiments of Faraday and of Thomas Andrews demonstra
ed the continuity of the gaseous and vapour states. Cf
Davy, 'Of the Corpuscular Theory',Works,II,p.330, from
his syllabus of lectures for 1802.

19. T.Thomson, System of Chemistry(1804)p.309; J.W.
Ritter, Nicholson's Journal,V(1803)255-6; W.H.
Wollaston, 'A Method of examining refractive and dis-
persive Powers',Phil.Trans.,XCII(1802)365-80. J.

Sowerby, A New Elucidation of Colours,original,
prismatic, and material;shewing their concordance in
three primitives, yellow, red, and blue: and the
means of producing,measuring, and mixing them:with
some observations on the accuracy of Sir Isaac Newton
(1809). The book is dedicated to the memory of
Newton, and the'observations'are not strictures. On
the Sowerby family, see the papers by J.B.MacDonald,
D.M.Simpkins, and R.J.Cleevely, Journal of the Society
for the Bibliography of Natural History,VI(1974)377-
559. On the colour systems, see the brief account in
W.Ostwald, The Colour Primer,ed.F.Birren(New York,1969).

20. T.Thomson, System of Chemistry(1804)p.311; W.T.
Brande, Manual of Chemistry(1819)p.58; by the edition
of 1845 Young and Fresnel were preferred; cf J.J.
Berzelius, Nicholson's Journal,XXXIV(1813)164-6. J.F.
Daniell, An Introduction to the Study of Chemical
Philosophy(1839)p.148; R.Kane, Elements of Chemistry,
(Dublin,1840)pp.18-22. See the papers on thermo-
chemistry by P.L.Dulong, G.H.Hess, and T.Andrews in
Phil.Mag.,XIX(3rd series,1841)19-25, 178-84. M.
Faraday, Chemical Manipulation(1827,reprint 1974).

21. W.Weldon, The Elements and General Laws of
Chemistry(1824)pp.44-5; cf W.Whewell's 'Demonstration
that all Matter is Heavy', in his Philosophy of the
Inductive Sciences(1847) II,pp.624-34.

22. J.M.Good, The Book of Nature(New York,1833)pp.x,
35, 44-5, 51ff., 325, 331.

23. J.M.Good, The Book of Nature(New York,1844)pp.62ff;
W.Herschel, Phil.Trans.,XCI(1801)265-318, 354-62;
Phil.Trans.,CI(1811)269-336; Phil.Trans.,CIV(1814)
248-84; quotation from CI(1811)331; XCI(1801)269,293.
See M.A.Hoskin, William Herschel and the Construction
of the Heavens(1963)pt.IV.

24. J.M.Lockyer, Solar Physics(1874)pp.435-524 for
reprints of Lockyer's papers; pp.1-12 for earlier
views. W.Parsons, 3rd Earl of Rosse, Scientific Papers
ed.C.Parsons(1926)pp.109-24 (a paper of 1850) eschewed
speculation, but as Agnes Clerke pointed out in her
Popular History of Astronomy, during the nineteenth
century (Edinburgh,1885)p.154, he was unusual in doing
so; see also her pt.II,chaps.2-5. See also R.Grant,
History of Physical Astronomy(1852)chaps.20 and 21;
and for more recent developments, J.D.North, The
Measure of the Universe(Oxford,1965); W.A.Miller and
W.Huggins, Phil.Trans.,CLIV(1864)413-35, 437-44. On
Lockyer, see A.J.Meadows, Science and Controversy
(1972).

25. T.S.Hunt, Chemical News,XLV(1882)74-6, 82-3; Isaac Newton's Papers and Letters on Natural Philosophy,ed.I.B.Cohen,Cambridge(1958)p.180. T. Carnelly, Chemical News,LIII(1886)157-9, 169-72, 183-6, 197-200. For speculation about inorganic evolution see chap.9 below; as an avowedly Newtonian approach to chemistry, see D.I.Mendeleev, 'An Attempt to apply Chemistry one of the Principles of Newton's Natural Philosophy', Royal Institution Library of Science,III, pp.540-59.

26. W.H.Brock, 'Studies in the History of Prout's Hypothesis', Annals of Science,XXV(1969)49-80, 127-37; we shall study in chap.6 below the attempts to test Prout's Hypothesis. See C.Hutton, Mathematical and Philosophical Dictionary(1795-6) II,p.82, under 'matter'. W.Prout, Bridgewater Treatise, 2nd edn(1834) pp.26, 35, and 40. The two views of crystals are those of R.J.Haüy and W.H.Wollaston; see D.C.Goodman, 'Problems in Crystallography in the Early Nineteenth Century',Ambix,XVI(1969)152-66: S.H.Mauskopf, 'Haüy's Model of Chemical Equivalence: Daltonian Doubts Exhumed',Ambix,XVII(1970)182-91; Wollaston's paper is in Phil.Trans.,CIII(1813)51-63.

27. W.Prout, Bridgewater Treatise,2nd edn.(1834)pp. 58-76; quotation from p.59.

28. W.Prout, Bridgewater Treatise,2nd edn.(1834)pp.83, 116, 152-3. One is tempted to infer an evolutionary connexion between Prout's sub-molecules, Crookes' sub-groups or meta-elements, and the modern idea of Iso-topes; but there may be none. I.Newton, Opticks, reprint of 4th edn.(New York,1952)pp.281-2.

29. Phil.Mag.,LIII(1819)142-6; contrast the view of Goethe, Theory of Colours,tr.C.E.Eastlake(1840)p.202; H.B.Jones, Life of Faraday, 2nd.edn.(1870)p.195; on Malus, see M.P.Crosland, The Society of Arcueil(1967) p.345.

30. H.Davy, Elements of Chemical Philosophy(1812)pp. 488-9; J.S.Mill, Autobiography,ed.H.Taylor, World's Classics edn.(1924)pp.135-6; C.Smith, 'Mechanical philosophy and the emergence of physics in Britain', Annals of Science,XXXIII(1976)1-30.

31. See D.M.Knight, 'Steps towards a dynamical chemistry',Ambix,XIV(1967)179-97; H.Davy, Works,IX, p.388; VIII,p.325; on nineteenth century atomism, see D.M.Knight, Atoms and Elements,2nd edn.(1970). C. Maclaurin, An Account of Sir Isaac Newton's Philo-sophical Discoveries(1748)p.20.

32. H.Davy, Works,VI,pp.245-56; ,quotation from pp. 245-6, 252. H.Hartley, Humphry Davy(1966)pp.134-6.

33. W.Morgan, Phil.Trans. Abridged,ed.C.Hutton et al., (1808) XV,pp.699-703; quotations from pp.703, 701. See J.Franklin, Narrative of a journey to the shores of the Polar Sea(1823)p.583, on aurorae.

34. M.Faraday, Experimental Researches in Electricity (1839) I,pp.486-96; Faraday's Diary,ed.T.Martin (1932-6) III,pp.14ff, 213.

35. J.J.Thomson, 'Cathode Rays', Royal Institution Library of Science,V,pp.36-49; for an aetherial rather than particulate view, see G.F.Fitzgerald, Nature,XXVIII(1888)446-8, his British Association Presidential Address.

5

Oxymuriatic Acid

In the last two chapters we have seen how the belief
that all matter was ultimately of one kind, received
support from theological arguments against materialism, and from the view that matter and light were
interconvertable. The first of these arguments led to
the doctrine that matter was passive, undifferentiated,
and inert, bodies owing all their properties to the
powers which arranged their particles. The second
might be held to imply that matter of whatever kind
could be got into a fourth state, of which there
would only be very few different kinds. Neither of
these arguments have very much to do with chemistry;
although as we saw the discoveries of Davy and his
contemporaries could be interpreted as giving support
to immaterialism, and the success of chemists in
liquefying previously 'permanent' gases and thus
showing that all substances could exist in all three
states might be held to support the theory of a fourth
state. But really both these chains of thought reach
back to the corpuscular philosophy of the seventeenth
century; and are therefore closer to physics than to
the mainstream of chemistry.

We must now therefore turn to chemical arguments
for the unity of matter. Some of these were metaphysical, depending upon chains of reasoning from

analogies and essentially unverifiable; others could
be falsified, and frequently were. The most famous
of this latter class was Prout's Hypothesis that the
atomic weights of all elements were integer multiples
of that of hydrogen. If this were true, it could be
conjectured that hydrogen was the only true element;
but atomic weight determinations showed that it was
not true. A sufficient number of atomic weights were
however close to integer multiples of that of hydrogen
to make the hypothesis fascinating even when apparently
disproved, for the world it opened up, composed of
only one kind of matter, was attractive. There were,
if this quantitative argument failed, qualitative
arguments to support the doctrine of the unity of
matter having a chemical basis; and it is to these
arguments that we shall first turn, looking especially
at the work of Humphry Davy.

Davy's greatest achievements were his advocacy of
the kinetic theory of heat, his studies of the oxides
of nitrogen, his electro-chemical researches culminating in the discovery of sodium and potassium, and his
proof that there was no oxygen in muriatic acid. A
formal link between all these lines of research is
that they all cast light on the nature of chemical
synthesis; apparently showing that neither heat nor
electricity is a substance, and that chemical elements
are not the bearers of properties. It seems likely
that these ideas were at the back of Davy's mind
throughout his researches, for we find them expressed
in his lectures although to his published papers he
often added disclaimers to the effect that his only
purpose was to bring forward facts. But broad and
general ideas about materialism and chemical synthesis
are not all that an opportunistic, ambitious and
nationalistic scientist needs in making his discoveries.

126

We must accordingly try to follow Davy as, reacting
to the ideas and discoveries of others and adhering
to his own general view of nature, he struggled to
understand the nature of muriatic acid. We shall be
looking at this piece of research with the question
of the unity of matter and the nature of the chemical
elements always before us; other viewpoints are no
doubt available, from which things would be seen
differently.

In a lecture of 1809 Davy said:

> That common air, nitrous oxide,
> nitrous gas, and one of the
> strongest known acids, should
> consist of two substances united
> in different proportions is a
> fact which, to be convinced of,
> requires the strongest evidence
> of experiment. Nitrous gas and
> nitrous acid again, though com-
> posed of the same elements, yet
> are themselves capable of combin-
> ation, as if they consisted of
> distinct materials. And these
> facts, combined with analogous
> facts relating to the compounds
> of hydrogen, carbon, and oxygen,
> render it probable that substances
> which we at present conceive to
> consist of different species of
> matter may ultimately be referred
> to different proportions of similar
> species; and in this way the
> science of the composition of
> bodies may be materially simplified.[1]

Davy went on to hope that investigation might confirm
his view that there was in all inflammable bodies a
principle of inflammability, and in all metals a
principle of metallization; for then a new, simple,
and grand philosophy would result:

> From the combination of different
> quantities of two or three species
> of ponderable matter, we might

conceive all the diversity of
material substances to owe their
constitution, and as the electrical energies of bodies are capable
of being measured, and as these
are correspondent to their chemical
attractions, so the laws of affinity
may be subjected to the forms of
the mathematical sciences, and the
possible results of new arrangements of matter become the objects
of calculation.

This last hope is only in our day being realised; the
passage is striking as showing a blending of the
Newtonian aspirations towards a chemistry based upon
quantitative laws of affinity, with a belief in
property-bearing principles descended from phlogiston.
What is striking is how in the next year Davy succeeded in dropping these principles, in the course of his
investigation of muriatic acid, without losing faith
in the unity of matter.

This story begins with the Swedish apothecary
Scheele who in 1774 had prepared a greenish gas by
the action of 'marine acid' on manganese dioxide.
His explanation of the reaction was that the manganese
attracted phlogiston from the acid; the gas was therefore dephlogisticated marine acid. This explanation
cannot be put into modern terms with precision; but
if we interpret phlogistication as reduction, then
Scheele's account is not incorrect in general terms.
Nevertheless, it was a casualty of the progress of
chemistry. Lavoisier's theory was that all acids
were compounds of oxygen, the generator of acids; and
marine or muriatic acid, our hydrogen chloride, must
be the oxide of an unknown basis, the muriatic
radical. The greenish gas was therefore a higher
oxide; and its new name was oxymuriatic acid.[2]

This gas was studied by Berthollet in 1785,[3] and

128

his researches led to the founding of the bleaching industry based upon it. He found that the gas was not itself acid; but confirmed that it formed muriates with metals without any gas being given off. The metal united, according to him, with part of the oxygen; and the salt was composed of the basic and acidic oxides like all others. Berthollet was of course unable to find the basis of muriatic acid; but with the coming of Volta's battery attempts were made to decompose muriatic acid electrically, though first with Franklinic electricity. One of the first to make these experiments was William Henry,[4] the friend of John Dalton.

Henry was afraid that Lavoisier's theory of acids, 'one of the most important parts of the new system of chemistry, must be regarded as incomplete, and liable to subversion, till the individual acids now alluded to have been resolved into their constituent principles'. He hoped that the electric fluid, an analytical agent more powerful than artificial heat, might bring to light the muriatic basis; but all his experiments produced negative results. He passed shocks through the dry muriatic acid gas in a glass tube over mercury, finding that the volume decreased but ascribing this to the presence of water. On sparking with oxygen he obtained oxymuriatic acid gas; in our view, he had decomposed the acid, but to him it appeared that oxygen had simply combined with the acid to yield the higher oxide. Other experiments with red-hot charcoal were also negative; and sparking the acid with 'carbonated hydrogen' gas was inconclusive. Henry concluded that it was very difficult to get his acid free from water, and that any apparent decomposition which had taken place had been due to the water, which had provided the oxygen to make the oxymuriatic acid.

In the same year one Pitchford set out to establish that oxygen and hydrogen were components of muriatic acid.[5] He followed a technique by which Fordyce had in the Philosophical Transactions for 1792 established that when metals dissolve in acids, they are oxidised by the oxygen of the water and not by that of the acid; this was in accordance with Lavoisier's view. Pitchford believed that muriatic acid was an exception to this rule. He dissolved one ounce of tin in four ounces of muriatic acid solution of a given specific gravity, precipitated the tin with caustic potash, and went on to neuralise the excess acid which required 1412 grains. But simply to neutralise four ounces of the acid, 1717 grains were required; while he allowed that some acid might have boiled away during the dissolution of the tin, the chief cause of the discrepancy seemed to be that some acid had been decomposed to oxidise the tin. But since the only products were hydrogen and muriate of tin, oxygen and hydrogen must be the components of muriatic acid. In an earlier chapter, we encountered a similar view in Girtanner; and Dalton also in his New System opted for the view that muriatic acid was an oxide of hydrogen.

Meanwhile on the Continent, men whose names are now equally forgotten were also at work. In 1805 Pacchiani wrote to Pignotti that he had proved by electrolysis that muriatic acid was water minus some oxygen. But attempts by Riffault to repeat these experiments, electrolysing water for thirty-four days with gold poles in glass apparatus, yielded no muriatic acid, and he concluded, in a paper translated into English in 1806, that Pacchiani's must have come from some animal or vegetable impurity, pure water yielding oxygen and hydrogen only.[6] This conclusion was verified by Davy's investigations published in his Bakerian Lecture of 1806; though it is worthy of note that as late as 1809 Davy found cause to complain that some authors still claimed to find nitrogen among the products of the electrolysis of water.

Despite Riffault's results, and in the same journal, J.Martin again urged the view that muriatic acid is an oxide of hydrogen.[7] He believed like Pitchford that metals dissolved in the acid were or could be oxidised by oxygen coming from the acid; which he prepared dry from salt and 'dense' sulphuric acid, and passed over red-hot iron wire in an earthenware tube in a furnace. Though everything was dry, muriate of iron and hydrogen were obtained in profusion; there could not have been enough water to generate the quantity of hydrogen that was found, nor to oxidise all the iron, and both these elements must therefore have come from the acid.

From our standpoint all these arguments seem wrong-headed because of their dependence upon Lavoisier's theory of acids. Henry displayed a certain scepticism towards it; and the posthumous publication of Black's Lectures on Chemistry provoked a rare challenge to this sensitive part of a powerful and valuable theory. In the Edinburgh Review the crucial questions were asked: 'How does it happen, that a body, admitted to be acid, should contain no oxygen? The Prussic acid is this body. And how comes it, that water, which is so highly oxygenated, has no properties of an acid?'[8] Lavoisier's remark that hydrogen was not an acidifiable base was simply a restatement of the problem in dignified language and not an explanation.

Very soon after the isolation of sodium and

potassium, Davy began work on this problem; and in his Bakerian Lecture of 1808 he reported that he had 'made a greater number of experiments upon this substance, than upon any of the other subjects of research that have been mentioned'. In his notebook we find an experiment in which he exposed dry potassium to the action of muriatic acid gas. According to the orthodox view, no hydrogen should have been evolved, for there would be none to be evolved; since Davy did not subscribe to the view that the acid was an oxide of hydrogen, he had to suppose that water was present despite his attempts to dry his apparatus and reagents, and he concluded that 'probably muriatic acid is incapable of acting in the gaseous state without water'. Attempts to make the dry acid by distilling muriates of tin and phosphorus, had failed unless water was present. We should remember that 'acid' at this time meant what we call anhydride; sulphuric acid would be SO_3 rather than H_2SO_4, if we may be anachronistic in using Berzelius' formulae, and salts were thought of as units composed of a basic oxide and an acid juxtaposed, MO,XO. To distil a salt might well yield the acid, therefore leaving the oxide; and it was disturbing that it did not, and that the acid could not even be decomposed by electricity. Davy even wrote to Dalton, probably about this time, that if these facts 'had been known 20 years ago, who would have given up phlogiston?'[9]

In 1809 Davy did some work on the evil-smelling substance hydrogen telluride; on 23 September he wrote to his friend J.G.Children: 'I doubt not you have found before this as I have done that the substance which we took for Sulphuretted Hydrogene is tellurated Hydrogene, very soluble in water, combinable with alkalies and earths, a substance affording another proof that Hydrogene is an oxide'.[10] This curious conclusion was arrived at because of the acid nature of hydrogen telluride. Tellurium was a metal, and while it might therefore like all metals contain hydrogen as a 'principle of metallization' it was certainly not an oxide; so the oxygen in the acid must come from the hydrogen. Davy had previously supposed that nitrogen might be an oxide, and his friends Allen and Pepys — who had written a long paper on respiration — had expressed the hope that he and Berzelius would succeed in analysing it. Now Davy proposed that hydrogen, nitrogen, and the oxides of notrogen were all oxides of ammonium. In nitrogen, the most stable, there was 'probably the most perfect equilibrium of affinities'.

But experiments during the next two months did not yield any evidence for this view, and in his Bakerian Lecture of 1809 he admitted that

> on the whole, the idea that ammonia
> is decomposed into hydrogene and
> nitrogene alone, by electricity, ...
> is in my opinion the most defensible
> view of the subject. But if ammonia
> be capable of decomposition into
> nitrogene and hydrogene, what, it
> will be asked, is the nature of the
> matter existing in the amalgam of
> ammonia? What is the metallic
> basis of the volatile alkali?

Davy is still in the grip of the doctrine of property-bearing principles, as were his contemporaries. Thus Berzelius considered nitrogen to be a metal; whereas Davy conjectured that either hydrogen was an element, and nitrogen an oxide; or that both were oxides of one metal, ammonium. These views seemed the best that could be formed on the antiphlogistic theory; but, he wrote: 'if these facts

133

concerning ammonia were to be reasoned upon,
independently of the other general phenomena of
chemical science, they perhaps might be more easily
explained on the notion of nitrogene being a basis,
which became alkaline, by combining with one portion
of hydrogene, and metallic, by combining with a
greater proportion'.[11] It is curious to find this
modern explanation thus labelled as 'phlogistic',
and to find phlogiston at all in the Philosophical
Transactions for 1810; and we should remember that
Davy had worked with Henry Cavendish, one of the
great 'phlogistic' chemists of the late eighteenth
century. But at this stage Davy concluded that to
say that a metal could be formed of two non-metallic
bodies was opposed to the tenour of chemical theory
in general.

He therefore returned to his theory that
hydrogen and nitrogen were oxides of ammonium; a
curious piece of looking-glass chemistry, where
removal of hydrogen is taken to be addition of
oxygen. Davy's next remarks were nevertheless
couched in phlogistic terms:

> If hydrogene according to an
> hypothesis to which I have often
> referred, be considered as the
> principle which gives inflammab-
> ility, and as the cause of
> metallization, then our list of
> simple substances will include
> oxygene, hydrogene, and unknown
> bases only; metals and inflamm-
> able solids will be compounds of
> these bases, with hydrogene; the
> earths, the fixed alkalies,
> metallic oxides, and the common
> acids, will be compounds of the
> same bases, with water.[12]

It is not obvious that a chemistry with these unknown bases would be simpler than one with the then-

accepted elements; but Davy presumably hoped that the number of bases would be smaller than the number of elements.

The evidence in favour of this view was firstly the 'inherent properties' of such elements as sulphur and tellurium which yield acidic hydrides. Secondly, many metals are obtained from their oxides by reduction with hydrogen, and may well therefore contain that element; this is indeed a curious survival of phlogiston. Thirdly, oxygen and hydrogen neutralise one another; all combustion may thus be a display of these 'natural energies', and furthermore the acidity of oxymuriatic acid, blunted by excess of oxygen, is increased when this excess is neutralised by hydrogen in the formation of muriatic acid. This is indeed an explanation of why oxymuriatic acid should be less acidic than muriatic, although it contained more oxygen, the generator of acidity; but in general these remarks of Davy's show him floundering and incoherent as he tries to follow the guide of Lavoisier's theory and of the phlogiston theory into a region where neither can without amendment fit the facts.

In his lecture, Davy had declared his support for the view of Gay-Lussac and Thenard that muriatic acid was composed of a body as yet unknown combined with water.[13] These two had in fact assembled most of the evidence which, some six months after the Bakerian Lecture, Davy was to reinterpret and use to overthrow Lavoisier's theory of acidity. They had themselves wondered whether it might not be simpler to regard oxymuriatic acid as an element; but had been persuaded by Berthollet to adhere to the received view. On combining potassium with muriatic acid, they found that just as much hydrogen was liberated as when an equal weight of potassium reacts with water. They

135

concluded that the acid contained, even when dried,
enough water to oxidise any metals that it could
attack. It was not until some time after Davy's
paper on chlorine that they came to accept his view;
in Davy and Gay-Lussac we see two of the foremost
chemists of the day working their way towards a more
satisfactory theory of acids, and by about 1815
arriving at similar theories by different routes.

Davy rapidly emerged from his phlogistic conjectures; Charles Daubeny, who became Professor of
Botany and Chemistry at Oxford, said of him that when
others took up his theories he dropped them. There
may be some truth in this — Davy always sought to be
original — but the situation in chemistry was like
that described by Alexander von Humboldt in geology:
'To boast of a stability of opinion ... is to boast
of an extreme indolence of mind; it is to remain
stationary amidst those who go forward'. Davy
adhered to belief in the unity of matter, and was
predisposed towards a Newtonian view of the world;
but in less general theories he was very flexible,
and was like Goethe opposed to systems and even to
consistency.[14]

In July 1810 Davy read to the Royal Society his
paper arguing for the elementary nature of oxymuriatic
acid. He combined tin, phosphorus, and ammonia with
the dry gas, and found no evidence of the formation
of oxides or of water; and similarly showed that no
water was produced when really dry oxymuriatic acid
and hydrogen were made to combine. He suggested that
any oxygen produced from oxymuriatic acid came in
fact from water:

> MM. Gay-Lussac and Thenard have
> proved by a copious collection of
> instances, that in the usual cases

where oxygene is procured from
oxymuriatic acid, water is always
present and muriatic acid gas is
formed; now, as it is shown that
oxymuriatic acid gas is converted
into muriatic acid gas, by combining with hydrogene, it is scarcely
possible to avoid the conclusion,
that the oxygene is derived from
the decomposition of water, and,
consequently, that the idea of the
existence of water in muriatic acid
gas, is hypothetical, depending upon
an assumption which has not yet been
proved — the existence of oxygene in
oxymuriatic acid gas.

Rather as an aside he demolished the French theory
that exothermic reactions were combustions: 'The
vivid combustion of bodies in oxymuriatic acid gas,
at first view, appears a reason why oxygene should be
admitted in it; but heat and light are merely results
of the intense agency of combination'. Davy's
previous achievement had been to introduce polarity
into chemistry; now he was to make oxygen share its
throne, and transform chemistry from a combustion
dominated science into one in which no reaction was
privileged.[15]

The arguments for accepting the elementary nature
of oxymuriatic acid despite the revision thus entailed
in the theory of acids — which seemed a deeply
entrenched part of chemical theory — were that it
provided a simpler account of the facts, and that it
suggested new analogies and opened paths to new discoveries which had previously seemed closed. But it
was to be some years before a coherent theory of
acids was available to replace Lavoisier's. Davy and
Gay-Lussac, the former after his studies of the
diamond, both came to the conclusion by about 1815
that acidity was a consequence of certain arrangements
of particles. We do not in Davy find a hydrogen

137

theory of acids, but there is in his paper of 1810 a
passage foreshadowing Laurent's:

> 'Hydrogene is disengaged from its
> oxymuriatic combination, in the
> same manner as one metal is dis-
> engaged by another, from similar
> combinations;' and 'When muriatic
> acid is acted on by mercury, or
> any other metal, the oxymuriatic
> acid is attracted from the hydro-
> gene, by the stronger affinity of
> the metal; and an oxymuriate,
> exactly similar to that formed by
> combustion, is produced.'[16]

The relation of muriates to oxymuriates continued to
puzzle some of Davy's contemporaries, who could not
believe that they were the same.

The name 'chlorine' first appeared in Davy's
Bakerian Lecture of 15 November 1810,[17] and is not
to be found in his notebook until the 24th. It was
chosen because, unlike both previous names for the
substance, it was free from theoretical presupposit-
ions, and could be used even of oxygen were after all
to be found in chlorine; Davy was always preoccupied
with finding a theory-free nomenclature. He credited
Scheele with the discovery of the elementary nature
of chlorine, remarking that Scheele's view had been
'obscured by terms derived from a vague and unfounded
general theory'. In a lecture of 1811 he was severe
towards the French school, and drew from the whole
episide the lesson of the danger of dogmatism in
science:

> The opinions of Berthollet have been
> received for nearly thirty years,
> and no part of modern chemistry has
> been considered as so firmly establish-
> ed, or so happily elucidated; but we
> shall see that it is entirely false —
> the baseless fabric of a vision ...
> oxymuriatic acid is not an acid, any

more than oxygen; but it becomes
an acid, like that substance, by
combining with inflammable matter
... The confidence of the French
enquirers closed for nearly a third
of a century this noble path of investigation, which I am convinced
will lead to many results of much
more importance than those which I
have endeavoured to exhibit to you.
Nothing is so fatal to the progress
of the human mind as to suppose
that our views of science are
ultimate; that there are no
mysteries in nature; that our
triumphs are complete, and that
there are no new worlds to conquer.

We can allow the discoverer his triumph, even when it
is accompanied by appeals to national prejudice; we
should notice that at this stage Davy seems to have
thought that chlorine was like oxygen a generator of
acids. As this class got larger, and his work on the
diamond showed the importance of arrangements, so he
generalised his theory of acidity.

At these triumphant lectures on chlorine was
Michael Faraday, who wrote to his friend Abbot in
August 1812: 'I would wish you not to be surprised if
the old theory of Phlogiston should be again adopted
as the true one tho I do not think it will entirely
set aside Lavoisier's but the Elements (Davy's
Elements, 1812) will inform you'. The Edinburgh
Review drew the same conclusion as Faraday, but
instead of rejoicing with Davy at the new worlds to
conquer the reviewer confessed to trepidation:

Should further inquiry justify this
conclusion, and explain away the
whole of the antiphlogistic doctrine
upon this point, we shall be obliged,
no doubt, to rejoice at the progress
of science; but our joy will be mixed
with trembling, when we reflect on
the precarious tenure by which chemical

> truth is holden; for it may,
> after this, be most reasonably
> doubted, whether the future dis-
> coveries shall not utterly destroy
> the merits of the later improve-
> ments in pneumatic chemistry, and
> bring us back to the doctrine of
> phlogiston, with the single except-
> ion of the positive levity of that
> body.[18]

These fears seem excessive, but serve to indicate how it seemed to contemporaries that Lavoisier's chemistry must stand or fall as a single edifice rather than as a number of distinct but related theories. In fact, Davy was going forwards rather than back, and there is no mention of Phlogiston in his Elements.

The new theory was slowly accepted.[19] In Nicholson's Journal Davy's younger brother John conducted a long-drawn-out battle with Murray, the author of a textbook, who supported the French view. The editor of the reviewing journal Retrospect of Discoveries regarded as crucial the reaction between dry ammonia and dry chlorine to produce ammonium chloride without any appearance of water; this seemed unlikely on the French view, while it followed as a matter of course on Davy's. On the Continent, the obscure Curaudau — who had tried to keep abreast of Davy's work on potassium — supported the new theory, and indeed claimed to have anticipated Davy. Among more celebrated chemists, Gay-Lussac and Thenard were after some time converted, but Dulong as late as 1813 still clung to the old terminology in a paper on nitrogen trichloride.[20] Berzelius held the anti-phlogistic view even longer, which might seem surprising in view of the wide agreement between him and Davy on matters of electrochemistry. But oxygen occupied an important place in his chemistry, and he was reluctant to displace it. The discovery of iodine convinced many doubters, for here was another substance like chlorine in its chemical properties, and it became harder to deny that there was a whole group of electronegative elements.

As late as 1815 William Henry of Manchester thought both views were equally good; and was worried about how chlorides formed in dry reactions were converted to muriates on solution in water. And the Edinburgh Review still voiced concern over Davy's work; his Bakerian Lecture of 1810 did not clearly demonstrate his views, and his nomenclature — much of which has not indeed survived — left much to be desired. It would have been well if, like Newton, he 'would attempt to confine himself within the ordinary limits of the language; for, really, it becomes a task of no small difficulty to trace his meaning through such neology'. The review also attacked him as being too tempted by hypotheses; which seems unjust since he had shown that his opponents' case rested upon hypotheses, or at least a general theoretical argument, whereas his was more closely based upon the observed chemical reactions of chlorine. To the charge of the reviewer that 'we observe him frequently prone to habits of less strict reasoning upon experiments invented with different objects', any discoverer in an experimental science would probably plead guilty unabashed.[21]

We have said that one of Davy's chief concerns was to reduce the number of chemical elements. Thus in 1809 he had declared that

> From the past progress of the human
> mind we have a right to reason con-
> cerning its future progress. And

141

on this ground a high degree of
perfection may be expected in
chemical philosophy. Whoever
compares the complication of the
systems which have been hitherto
adopted, and the multitude, as
it were, of insignificant elements
with the usual simplicity and
grandeur of nature, will surely
not adopt the opinion, that the
highest methods of our science are
already attained; or that events
so harmonious as those of the external world should depend upon
such complex and various combinations of numerous and different
materials.[22]

It is therefore curious to find him in his research
upon chlorine actually proving that a substance
believed to be compound was beyond reasonable doubt
elementary. In the event, this work on chlorine did
not weaken his general belief, but on the contrary
provided fresh analogies which seemed to support it.
Such general beliefs are compatible with a wide range
of experimental evidence, and in the day-to-day
business of the laboratory hypotheses on a lower level
of generality are required.

With the investigations of chlorine, Davy's
period of greatest activity in fundamental research
drew to a close, although his work on the safety lamp
and flame was still to be done. In 1812 he married
a wealthy widow, Mrs Apreece, and gave up his lecturing at the Royal Institution; writing to his friend
Clayfield: 'Having given up lecturing, I shall be able
to devote my whole time to the pursuit of discovery.'[23]
This was not the way things turned out; but in Davy's
Elements, 1812, which appeared soon after the investigation into chlorine and contains chiefly accounts of
those regions of the science of chemistry in which
Davy had himself worked, we find passages which link

142

his new view of chemical synthesis with his speculations about the nature of the elements. The book was
praised for its immaterialism, notably in the
Philosophical Magazine where the reviewer applauded
Davy's explanation of the heat evolved in chemical
reactions as

> a much more simple and more natural
> exposition of the actions of matter,
> than the supposed legerdemain
> operations of phlogiston or oxygen,
> which have been so confidently main-
> tained by theorists. Many hitherto
> undecompounded bodies [Davy's term
> for elements], which cannot easily
> be supposed to contain oxygen, pro-
> duce heat and light by their mutual
> chemical action.

Early in the book, Davy's views about the
elements appear.[24] As in his lectures, he suggested
that complexity was a feature of the early history of
a science: this idea came from the history of
astronomy, where the clumsy machinery of spheres and
epicycles had been superseded by Kepler's ellipses
and Newton's law of gravity; Davy and his contemporaries were constantly on the watch for the Kepler and
Newton of chemistry, and were not above putting their
own names forward as candidates for these honours.
This astronomical paradigm made it evident that

> it is contrary to the usual order
> of things, that events so harmonious
> as those of the system of the earth,
> should depend on such diversified
> agents, as are supposed to exist in
> our artificial arrangements; and
> there is reason to anticipate a
> great reduction in the number of
> the undecompounded bodies, and to
> expect that the analogies of nature
> will be found conformable to those
> of art.

While the idea of progress from complexity to
simplicity was derived from the history of astronomy,
the contrast between natural and artificial arrangements was a feature of contemporary botany of which
more must be made when we come to examine chemical
taxonomy, in the schemes of classification leading
up to Mendeleev's Periodic Table.

Davy had arguments rather more cogent than
analogies from neighbouring sciences, the most important of which was that the sciences were a unity.
Unlike Whewell and Hegel, for whom the boundaries
between the sciences were natural,[25] and unlike those
chemists who were suspicious of using physical
evidence in their science, Davy was a believer in the
intimate connection between the sciences. A theory
of matter based upon chemical considerations alone
would not have been adequate for one who believed
that 'the more the phenomena of nature are studied,
the more distinct their connection appears, the more
simple their causes, the more magnificent their
design, and the more wonderful the wisdom and power
of their author'. It is perhaps a pity that Davy did
not survive into the 1830s, when he might have given
us a Bridgewater Treatise on chemistry and natural
theology more interesting than Prout's; and no doubt
it was for passages like this that his Elements was
commended in the Philosophical Magazine to clergymen.

The union of chemistry and electricity brought
about largely by Davy's researches is an example of
this connection, for the phenomena of two sciences
were shown to be the result of the action of one
power. The reviewer[26] welcomed Davy's electro-
chemical views, remarking that there was nevertheless
as yet no wholly-satisfactory theory available:
'Although the theoretical error of the present age may

144

be that of attempting to simplify everything too far, and to refer numerous phaenomena to one general principle; yet it would appear that some simple law, to which all the others are subordinate or subserviant, is still a desideratum in the science'. After Davy's death, the great French chemist J.B.Dumas wrote of his passion for a unified theory, referring to his poetry, his imagination, his pantheism, and his interest in the alchemists. We need not take this too seriously, but we may note Whewell's remarks on Davy's introduction of polarity into chemistry: according to him, Davy's correct but very general ideas had to be made concrete by the addition of hypotheses which then, in a dialectical progress,had to be refined away by Faraday.

Davy's contemporaries indeed spoke of his boldness and his imaginative power.[27] Coleridge and Southey thought that he might have made a great poet, J.A.Paris (his biographer) and Thomas Thomson contrasted the leaps of Davy's mind and the rapidity of his experiments with the exact and scrupulous methods of Wollaston. Sir Benjamin Brodie, the physiologist, and protégé of Davy, who like him achieved the Presidency of the Royal Society and a baronetcy, in his recollections of Davy refers to this precipitancy and also to the caution over theorizing that was also a feature of the man; who had written after the reception of his 'phosoxygen' paper of 1799 'I began the pursuit of chemistry by speculations and theories; more mature reflection convinced me of my errors, of the limitations of our powers, the dangers of false generalisations, and of the difficulty of forming true ones'. By 1802 he had come to stress the importance of analogy in the sciences, which were 'classes of different facts associated together by analogy'.

Theories were abstract and artificial arrangements by which 'the business of thinking is materially expedited; and the power of calculating new relations of facts rendered simple and easy'. His view of science was nevertheless at this date exalted, and he declared in another lecture that it should be pursued for its 'own sake rather than with any idea of profit or interest'; this passage should not be taken in isolation, but in relation to the Royal Institution's transition from a technological institute to something more like a centre of pure research.

In 1807 Davy expanded his view of analogy in science, of which it was the 'governing spirit', connecting, arranging and employing related or similar facts. With justly-chosen analogies, men of science had since the days of Bacon and Newton 'examined with reverence and awe, the substantial majesty of nature. Discovery has not visited them and disappeared again, like the flashes of lightening amidst the darkness of night; but it has slowly and quietly advanced, as the mild lustre of the morning, promising a glorious day'. Not all the analogies chosen even by professed disciples of Bacon and Newton were happy; dreams of reducing the biological sciences to mechanics, pneumatics, or electricity were bound soon to pass away. As he argued in a lecture following the research on chlorine, he was not against all theory but only against premature development of it; 'against hasty generalisations which, pretending only to predict what will hereafter occur, fix, as it were, boundaries to the empire of science, and bring to the standard of a weak and diseased fancy the august dominion of nature'. In a long English tradition, he was against systems and dogmas, but did not deny the economical and heuristic value of hypotheses.[28]

We saw this in his taking up and dropping of 'phlogistic' views as he worked his way towards an understanding of chlorine; but we find in his arguments from analogy for the unity of matter something a little more firmly based. Dogmas resulted from the use of fancy; if Davy was usingthis word like his friend Coleridge did then in contrast to it one could set reason and imagination, powers which revealed the true analogies of nature. The exercise of these powers demanded a certain humility and natural piety, and a patient following of the methods of Bacon and Newton. The nature of the chemical elements could only be reasoned on by analogy, for by definition they could not be analysed. The number of known elements grew rapidly in the early nineteenth century, and it became clear that they fell into families. Davy isolated sodium and potassium; he and Berzelius prepared the alkaline earth metals; he identified chlorine, and, in a race with Gay-Lussac, iodine, as elements; and his friends Wollaston and Tennant discovered metals of the platinum group. The analogies between the properties of the elements in the various groups seemed evidence of complex nature; why should the world be composed of irreducibly different building-blocks as similar to each other as are sodium and potassium? It was simpler to suppose that these bodies were compound, and the ammonium radical seemed 'a kind of type of the composition of the metals'.[29] This argument, that elements and compound radicals behave similarly and that since the latter are known to have a complex structure it is probable that the former are also compound, enjoyed wide popularity during the nineteenth century; thus

147

sixty years after Davy's remarks we find Edmund Mills arguing for the complexity of chlorine because of its analogies with the benzoyl radical. We shall pursue this story in a later chapter.

Davy organised the metals into a series which — following an eighteenth-century idea — he called 'one chain of natural bodies'; the non-metals did not fit in so happily.[30] He suggested that all metals might be compounds of hydrogen; or that all bodies might be formed of the same matter in different states or different arrangements (he instanced ice, water, and steam); or that oxygen and hydrogen might be the same body in different electrical states; or that all bodies might be made up from two or three elements in different proportions. Thus chlorine might well be a compound of oxygen and an unknown basis; its colour, high equivalent weight, and solubility in water all favoured this view. These curious queries give rise to the suspicion that Davy was ensuring that he could not be wrong by putting down in vague terms all possible views of the question. All that he is committed to is the general belief that the elements are not composed of true indivisible atoms. Some of his suggestions are distinctly odd; that potassium might contain hydrogen was an idea which he had re-futed by experiment when it had been proposed by Gay-Lussac and others soon after Davy had discovered it; and he had also just refuted the generally-held belief that chlorine was an oxide. The point being made is that potassium is no more an hydride than is any other metal; but all might perhaps be shown to contain hydrogen, more strongly combined than in ordinary com-pounds, and maybe a simpler chemistry would result therefrom. Similarly, chlorine was no more an oxide than other non-metals; but all might be found to be very stable compounds of oxygen. That oxygen and hydrogen were differently-charged forms of water had been Ritter's view, and Davy had toyed with it during his 'phlogistic' period; but by about 1810 the only really strong candidate was the revived corpuscularian view that the same matter in different arrangements might constitute all bodies, and it was indeed to this view that Davy turned after his work on the diamond.

Some of these suggestions depend upon property-bearing elements, and others do not; some must have seemed more convincing to his readers than others. We noted that Dumas referred to Davy's reading of the alchemists; Davy in a lecture was careful to separate philosophical reasons, based upon analogy, from the empirical doctrines of alchemy. Chemistry was in the same position as sixteenth-century astronomy, still awaiting its Newton to 'show the magnificence and simplicity of nature'.[31] Even now, it was not uninstructive or uninteresting despite its unmathematical nature; and at the end of another lecture he urged the chemist

> to be bold in pursuit. He must not
> consider things as impractible,
> merely because they have not yet
> been effected. He must not regard
> them as unreasonable because they do
> not coincide with popular opinion.
> He must recollect, how contrary
> knowledge sometimes is to what appears
> to be experience. Our senses seem to
> prove to us, that the earth is a plane
> surface and at rest; — our science
> informs us that it is round and revolves.

These passages of Davy's writings and lectures are of some interest in that they point towards Prout's Hypothesis;[32] in his conjecture that all elements are built up of hydrogen Prout seems to

149

have been building upon Davy's vague suggestions,
and as we shall see later defenders of the unity of
matter happily followed Davy's advice not to take
experience — or experiment — too seriously when it
conflicted with what was felt to be knowledge.

NOTES

1. H.Davy, Works,VIII,pp.323-5.

2. C.W.Scheele, 'On Manganese', Alembic Club Reprints,XIII(1897) 5-10; A.L.Lavoisier, Elements of
Chemistry(1790)pp.71-3.

3. C.Berthollet, 'Memoir on Dephlotisticated Marine
Acid', Alembic Club Reprints,XIII(1897) 11-31; H.E.
Le Grand, 'Ideas on the Composition of Muriatic Acid
and their relevance to the oxygen theory of acidity',
Annals of Science,XXXI(1974) 213-26; 'The "Conversion"
of C.L.Berthollet to Lavoisier's Chemistry', Ambix,
XXII(1975) 58-70.

4. W.Henry, Phil.Trans.Abridged,ed.C.Hutton et al.
(1809)XVIII,pp.641-9; the article appeared in 1800.
The quotation is from Henry's paper in Nicholson's
Journal,IV(4th series,1800) 209. On the Henrys, see
the papers by W.V. and K.R.Farrar and E.L.Scott,
Ambix,XX(1973) 183-208; XXI(1974) 179-228; XXII(1975)
186-204; XXIII(1976)27-52; XXIV(1977)1-26.

5. J.Pitchford, Nicholson's Journal,4° series,IV
(1800) 374; G.Fordyce, Phil.Trans.,LXXXII(1792) 374-
82. For Girtanner's experiments, see above,chap.2,
note 17; J.Dalton, A New System of Chemical Philosophy
pt.II(1810)p.293.

6. J.R.D.Riffault, Journal de Physique,LXI,p.281; a
slightly condensed translation appeared in Nicholson's
Journal,XIII(1806) 137; see also Phil.Mag.,XXVIII
(1807) 55-65; H.Davy, Phil.Trans.,XCVII(1807) 1-56;
C(1810) 40.

7. J.Martin, Nicholson's Journal,XIII(1806) 237.

8. Edinburgh Review,V(1803) 25. But note C.L.
Berthollet, Essay on Chemical Statics,tr.S.Lambert
(1804)II,p.8: 'In fact it is carrying the limits of
analogy too far to infer that all acidity, even that
of the muriatic, fluoric, and boracic acids, arises
from oxigen, because it gives acidity to a great
number of substances. Sulphuretted hidrogen, which
really possesses the properties of an acid, proves,

that acidity is not, in all cases, owing to oxigen'.
One must not overestimate dogmatic devotion to paradigms.

9. H.Davy, Phil.Trans.,XCIX(1809) 91; Laboratory
Notebook,29 December 1808; letter to Dalton, at the
Royal Institution,1809.

10. H.Davy, Phil.Trans.,C(1810) 27-31; letter to
J.G.Children, 23 September 1809, at the British
Museum; on Children, see W.C.Smith, Bulletin B.M.
(Nat.History) Historical Series,III(1969) 250-1;
W.Allen and W.H.Pepys, 'On Respiration', Phil.Trans.,
XCIX(1809) 428.

11. H.Davy, Phil.Trans.,C(1810) 55; J.J.Berzelius,
Nicholson's Journal,XXXIV(1813) 240; and Annals of
Philosophy,II(1813) 276-84, 357-68,nb.pp364-7; cf
H.Davy, Phil.Trans.,XCVIII(1808) 362; XCIX(1809) 55-6;
C(1810) 38-44. 68-73.

12. H.Davy, Phil.Trans.,C(1810) 69.

13. H.Davy, Phil.Trans.,C(1810) 67; J.L.Gay-Lussac
and L.J.Thenard, Memoires d'Arcueil,II(1809) 339, 342,
357; parts of this paper are translated in Alembic
Club Reprints,XII(1897) 34-48; see M.P.Crosland, The
Society of Arcueil(1967)pp.309-10, and his 'Lavoisier's
Theory of Acidity',Isis,LXIV(1973) 306-25.

14. J.Davy, Memoirs of H.Davy(1836)II,p.125; and
Daubeny's remarks in Phil.Mag.,VIII(3rd series,1836)
249. Cf A.von Humboldt, Personal Narrative,tr.T.Ross
(1852-3)III,p.377; H.Maine, Ancient Law,Everyman edn
(1965)p.41, on the 'doubtful virtue' of consistency;
Goethe:Conversations and Encounters,ed. and tr. D.Luke
and R.Pick(1966)pp.14, 218-9.

15. H.Davy, Phil.Trans.,C(1810) 230-57; parts of this
and other papers by Davy on chlorine appear in Alembic
Club Reprints,IX(1894); quotations are from pp.236
and 239; J.M.Good, The Book of Nature (New York,1833)
p.48; M.P.Crosland,Gay-Lussac,(Cambridge,in press).

16. H.Davy, Phil.Trans.,C(1810) 240,238; A.Laurent,
Chemical Method,tr.W.Odling(1855)p.39.

17. H.Davy, Phil.Trans.,CXI(1811) 32; Elements of
Chemical Philosophy, (1812)p.vii; Phil.Trans.,C(1810)
237; Works,VIII,pp.313-8; For criticism of his neologisms, see Edinburgh Review,XXIII(1815) 492. J.Z.
Fullmer, Davy's Published Works (Cambridge,Mass.,1969)
pp.61-3; M.P.Crosland, The Language of Chemistry(1962)
pp.221-2.

18. L.P.Williams, 'Faraday's Education in Science',

Isis,LI(1960) 515; and (ed.) Selected Correspondence of Michael Faraday,2 vols(Cambridge,1971)I,16; Davy disavowed his phlogistic speculations in Phil.Trans. C(1810) 246,ftnt.; Edinburgh Review,XVII(1811) 405.

19. For the ding-dong battle between John Davy and Murray, see Nicholson's Journal,XXVIII-XXXIV(1811-3); John Davy pursued a career of some distinction in science and medicine, and went on to become Inspector General of military hospitals. Retrospect of Discoveries,XXXII(1812) 48,130; Curaudau was an industrial chemist — see W.Scoresby, An Account of the Arctic Regions and of the Whale-Fishery(Edinburgh, 1820)II,p.431.

20. J.B.Dulong, Memoires d'Arcueil,III(1813) 48; for Gay-Lussac and Thenard, see Alembic Club Reprints, XIII(1897) 49; M.F.Crosland, The Society of Arcueil (1967)pp.309-10. J.J.Berzelius, Annals of Philosophy VI(1815) 211; Davy's reply, Quarterly Journal of Science and Arts,I(1816) 68. On iodine, see H.Davy, Phil.Trans.,CXIV(1814) 74-93; and P.Collins, 'Humphry Davy and Heterogeneous Catalysis',Ambix,XXII(1975) 205-17, 208; and J.Z.Fullmer,'Davy's Priority in the Iodine Dispute',Ambix,XXII(1975) 39-51.

21. W.Henry, The Elements of Experimental Chemistry, 7th edn(1815) I,p.398; The Edinburgh Review,XVIII (1811) 470,479.

22. H.Davy, Works,VIII,pp.325-6; there is a version of Ockham's Razor which goes 'frustra fit per plura, quod fieri potest per pauciora', and this passage could be taken as an application of this principle in chemistry.

23. J.Davy, Memoirs of H.Davy,I,p.143; H.Davy, Elements of Chemical Philosophy (1812); Phil.Mag., XL(1812) 145,297,434.

24. H.Davy, Works,IV,p.41; VII,p.97-8.

25. See Hegel's Philosophy of Nature,ed. and tr.M.J. Petry (1971); W.Whewell, Philosophy of the Inductive Sciences,2nd edn(1847) especially I,pp.78-81; contrast Baden Powell, Essays(1855)pp.40-81; H.Davy, Works,IV,p.41; T.H.Levere, Affinity and Matter (Oxford,1971)pp.66-7.

26. Phil.Mag.,XL(1812) 307 J.B.Dumas, Lecons sur la Philosophie Chimique,Paris,1837,IXme lecon; see D.M. Knight, 'The Scientist as Sage', Studies in Romanticism,VI(1967) 65-88.

27. H.Hartley, Humphry Davy(1966)pp.22-5,142-3; J.A. Paris, The Life of Sir H.Davy(1831)I,p.145, II,p.402;

The Works of Sir B.C.Brodie, (1865) I, p.57 — J.Davy,
Memoirs of H.Davy(1836) I, p.147, records a legacy
from Davy to Brodie; see also T.Thomson, History of
Chemistry, 2nd edn(1830-1) II, pp.246,256; J.Davy,
Memoirs of H.Davy(1836) I, p.80; H.Davy, Works, II.p.329;
VIII, p.164. See T.Martin, 'Early Years at the Royal
Institution', British Journal for the History of
Science, II(1964)99-115; The Collected Works of Count
Rumford, ed.S.C.Brown, (5 vols., Cambridge, Mass., 1968-
70) gives a good idea of the mixture of pure and
applied science which the managers of the Royal
Institution no doubt expected from Davy; and see M.
Berman, 'The Early Years of the Royal Institution',
Science Studies, II(1972)205-40.

28. H.Davy, Works, VIII, pp.167,179,285-6,317.

29. H.Davy, Elements of Chemical Philosophy(1812)
p.481; E.Mills, Phil.Mag., XLII(4th series,1871)125;
cf Davy, Works, VIII, p.323.

30. H.Davy, Elements of Chemical Philosophy(1812)pp.
478-89; J.Davy, Memoirs of H.Davy(1836) I, p.467; R.
Siegfried, 'Sir Humphry Davy ... on the diamond',
Isis, LVII(1966)325-35.

31. H.Davy, Works, VIII, pp.261,326,331.

32. Anon[W.Prout], Annals of Philosophy, VI(1815)321-
30; VII(1816)111-3; H.Davy, Elements of Chemical
Philosophy(1812)p.481.

6

The First Principles of Chemistry

Davy's arguments from analogy on the nature of the
elements were not untypical of the period, though
not all chemists would have agreed that this was a
hopeful line of enquiry. In this chapter and the two
which follow it, we shall look at speculations on the
subject between about 1810 and 1870. This period includes the chemical work of Faraday and many other
eminent scientists and also the attempt of William
Prout to give a quantitative form to the theory of
the unity of matter. We shall also look at some
authors whose work, for reasons good or bad, is no
longer widely remembered.

Samuel Parkes, whose Chemical Catechism had by
1822 reached its tenth edition and become a standard
work for children, wrote that 'since the truly
interesting and very important discoveries of Sir
Humphry Davy, it is scarcely possible to say what
substances are not compound bodies'.[1] Parkes did not
pursue the matter further. In the more advanced textbook of William Henry we find the remark that chemists
are probably still far from a knowledge of the true
elements of matter, and hence that any classification
which began with elements was liable to frequent and
fundamental change. Similar statements can be found
in other textbooks, some authors being content with

this state of affairs while others deplored it.[2]

Non-chemists also interested themselves in this
problem.[3] Dugald Stewart, an influential member of
the Scottish 'common-sense' school of philosophers, saw
the eminent chemist Richard Chenevix's unfortunate
attempt to prove palladium an alloy of platinum.
Chenevix — the Copley Medallist of 1803 — had urged
chemists to simplify their science by analysing
alleged elements instead of adding to the number, and
believed that he had succeeded in his own programme.
Stewart judiciously discussed the new elements:

> the names that support these
> discoveries are respectable, and the
> experiments decisive. If we do not
> give our assent to them, no single
> proposition in chemistry can for a
> moment stand. But whether all these
> are really simple substances, or com-
> pounds not yet resolved into their
> elements, is what the authors them-
> selves cannot possibly assert; nor
> would it, in the least, diminish the
> merit of their observations, if future
> experiments should prove them to
> have been mistaken, as to the
> simplicity of these substances.
> This remark should not be confined to
> later discoveries, it may as justly
> be applied to those earths and metals
> with which we have been long
> acquainted.

Stewart was not simply making a logical point, that
elements defined in what we have called a negative-
empirical way as the limits of analysis, will be unstable as new methods of analysis are perfected. Two
pages further on he refers to those who wished to
reduce the number of elements because of their conviction that the world was a simple place:

> It required nothing less than the
> united splendour of the discoveries
> brought to light by the new chemical

156

school to tear the minds of men
from the pursuit of a simple and
primary element; a pursuit renewed
in every age with an indefatigable
perseverence, and always renewed in
vain. With what feelings would the
physiologists of former times have
looked down on the chemists of the
present age, whose timid and circum-
scribed system admits nearly forty
different principles in the composition of bodies! What a subject
for ridicule would the new nomenclature have afforded to an alchemist.

Hartley with his doctrine that all mental phenomena
were to be ascribed to vibrations and associations
was an 'alchemist of the mind'; but Stewart seems to
have thought that the days of this kind of alchemy in
chemistry were over. By the end of the century the
number of 'principles' had doubled; but the new alchemy of Rutherford was soon to appear.

Another philosopher who interested himself in
the problem and remained unconvinced by the reduction-
ists was Auguste Comte.[4] He was suspicious of atomic
theories, and of any attempt to connect physics with
chemistry; regarding as very chimerical any attempt
at universal explanations by a single law. Chemistry
was at present, for Comte, an 'erudition' rather than
a 'science', and chemists should study taxonomy; dualism should he believed be made a constant and necessary principle of classifying compounds, a principle to
which the doctrine of definite proportions should be
sacrificed if necessary. Chemistry could not be
reduced to mathematics; this would be 'profoundly ir-
rational', 'contrary to the nature of the phenomena',
and 'would occasion vast and rapid retrogradation, by
substituting vague conceptions for positive ideas,
and a new algebraic verbiage for a laborious investigation of facts'. To the founder of Positivism, such

analogical arguments as those of Davy with their
metaphysical foundation in a belief in the simplicity
of nature could not but be anathema. We have no
right, he declared, to assume axioms of that kind
before beginning work in science; chemistry had, on
his view, been an absolute prey of hypotheses, and he
found even the doctrine of affinity 'metaphysical'.
Where Cuvier had argued that the unequal distribution
of the various elements indicated that they were not
all of equal status, Comte believed that those rare
in the Earth's crust may be abundant in the core.

   We have already met the physician John Mason
Good, who lectured from 1810 at the Surrey Institution, and had something to say about atomism and about
the nature of the elements.[5] He was opposed to the
point centres of force of Boscovich since the points
must have weight, for otherwise they would be non-
entities; and if they have weight, then they are
atoms. The search for elements had been in vain; fire,
earth, air, and water had been replaced by phlogiston,
acids and alkalies, and then by oxygen, nitrogen,
hydrogen, and carbon; but all had 'been proved, or
reasonably suspected, to be compounds of other substances, that may yet, perhaps, be detected to be
compounds of something beyond'. For Good, there were
no elements; there were only the atoms of Epicurus:
'one substance is just as much entitled to the
character of a constituent element as another, — and
... everything equally proceeds from and in turn is
resolved into the primitive and invisible atoms or
principles of matter'. We saw that Good's natural
theology had a flavour of the seventeenth century, and
so did his theory of matter. The atoms must compose
substances on some hierarchical principle; and transmutations would simply represent chemical change at a

deeper level. Indeed, Good's world seems to have been in flux, with such transmutations occurring all the time; this is not surprising in one whose chief interests were in the biological sciences, for the belief was widespread that organisms could transmute. Nevertheless, he was not despairing of chemistry, for 'although we have not been able to follow-up the Proteus-form of matter into its deepest recesses, and fix it in its last shape and character — we have succeeded in developing many of its most important laws'.

The Surrey Institution closed down after living for some years on its capital, its copies of Nicholson's Journal found their way into the Bodleian Library, and we hear no more of Good's views on chemistry. But in 1815 and 1816 in two short anonymous papers in Annals of Philosophy Prout published his hypothesis.[6] The papers are a surprise, for only the second contains a clear statement of the conjecture that all elements are made up of hydrogen. The papers seem to have come out of the controversy between Dalton and Gay-Lussac over whether gases combine in simple proportions by weight or by volume. Prout found an agreement between the atomic weights and specific gravities of a number of gases, using 'atom' to cover both our 'atom' and 'molecule' as was then general in England. He was also struck by the 'near approach to round numbers of many of the weights of the atoms'. His conclusions from his table of results were 'that all the elementary numbers, hydrogen being considered as 1, are divisible by 4, except carbon, azote, and barytium, and these are divisible by 2, appearing to indicate that they are modified by a higher power than that of unity or hydrogen. Is the number 16, or oxygen? And are all

substances compounded of these two elements?' Even
as an undergraduate, he had believed in the unity of
matter; but the quantitative basis for the belief
was novel, and was in tune with the spirit of
chemistry as it was being transformed by Dalton's
theory.

The first paper could have been taken as an
argument for using H=1 rather than O=10 as the basis
for a scale of atomic weights. But in the second
paper, Prout made his position clear; first arguing
for what later chemists would call a 'one-volume
system', that is, for taking the volume of hydrogen
as equal to the atom, or, in our terms, assuming that
the molecule of hydrogen contains one atom so that
the molecular weight of hydrogen is unity. As Prout
pointed out, on such a system the specific gravities
of gaseous elements will, as a rule, coincide with
their atomic weights. But this convenience was not
merely artificial, for Prout took up Davy's remark
that hydrogen perhaps approached nearly to what we
might expect a real element to be like:

> If the views we have ventured to
> advance be correct, we may also
> consider the πρωτη ὑλη of the
> ancients to be realised in hydro-
> gen; an opinion, by the by, not
> altogether new. If we actually
> consider this to be the case, and
> further consider the specific
> gravities of bodies in their gas-
> eous state to represent the number
> of volumes condensed into one; or,
> in other words, the number of the
> absolute weight of a single volume
> of the first matter (πρωτη ὑλη) which
> they contain, which is extremely
> probable, multiples in weight must
> always indicate multiples in volume,
> and vice-versa; and the specific
> gravities, or absolute weights of

all bodies in a gaseous state, must
be multiples of the specific gravity
or absolute weight of the first
matter (πρωτη ὑλη) , because all
bodies in a gaseous state which unite
with one another unite with reference
to their volume.

We notice here the stress on condensation, on density
as the measure of the quantity of matter in a given
volume, which is characteristic of the speculations
of eighteenth-century Newtonians; we find a not dissimilar idea, though divested of its corpuscularian
basis, in the view of chemical synthesis taken by
Sir Benjamin Brodie the younger in the 1860s — we
shall return to him and his chemical calculus in due
course, in a later chapter.

With this concern for the relations between
volumes and atoms it is not surprising that a few
years later Prout independently hit upon the hypothesis Avogadro had proposed in 1811, that equal
volumes of gases contain, under the same conditions,
equal numbers of particles.[7] Naturally his two papers,
in the chemical atmosphere of the day and appearing
in a lively journal, aroused considerable attention;
what is perhaps surprising is that Prout made very
little further effort to justify his hypothesis.
Measurements at what were then the limits of accuracy
were not easy to make; and this task fell on the will-
ing shoulders of Thomas Thomson, in whose journal the
papers had appeared and who later divulged their
authorship. Thomson was merely the first of a number
of eminent chemists who became converts to Prout's
hypothesis in one form or another during the nineteenth
century, as the idea of prime matter continued to
exert its fascination.

Thomas Thomson was the populariser both of

161

Dalton's atomic theory and of Prout's hypothesis, and is therefore the great exception to the generalisation that those who believed the elements to be complex rejected Daltonian atomism. It was in Thomson's System of Chemistry that Dalton's theory was first brought to the attention of the world; and the papers of Thomson and of Wollaston on the oxalates, with their evidence for definite and multiple proportions, gave the theory a firm basis in chemical analysis. In his journal, Annals of Philosophy, Thomson took every opportunity of popularising Daltonian atomism;he followed Wollaston and Berzelius in taking oxygen as his standard in the atomic weights which he began energetically to determine, using O=1 as the basis of his scale. Berzelius had, in an article in Thomson's journal, proposed the notation which we still use for indicating atoms; although for Berzelius at this period they stood not for atoms, which were hypothetical, but for volumes; oxygen being the standard for specific gravities of gases, expressed on a scale relative to O=100.[8]

Thomson was not a great admirer of Davy; his was a methodical mind. Thus in 1813 he wrote that Davy had 'embraced the Daltonian theory with some modifications and alterations of terms; but his notions are not so perspicuous as those of Mr Dalton, and they do not appear to me so agreeable to the principles of sound philosophy'.[9] Davy's championship of the Higgins's claims to have anticipated Dalton must have set Thomson against him; and in 1814 Thomson, after dealing with the Higginsites remarked that neither Davy nor Berzelius were atomists, and that Davy had indeed written 'two very violent and I

162

think indecorous notes against it' although his theory embraced much of Dalton's.

Up to 1814, Thomson appears as a strict Daltonian atomist, interested in determining atomic weights; and in that year he announced his coversion to Davy's views on chlorine,[10] which earlier had seemed to him to threaten the whole theory of acidity. He felt that one of the most important effects of the atomic theory had been the impulse it had given to accurate quantitative experimental work; a sentiment which Davy was to echo a decade later when it fell to him as President of the Royal Society to make the speech on the award of a Royal Medal to Dalton. In 1816 Thomson, fired by Prout's hypothesis, began the work, a strange amalgam of speculation and sober analysis, which entitles him to his place in this story.

Every year he published, as Berzelius was later to do, a sketch of improvements in the sciences, and especially in chemistry, during the previous year. In reviewing 1815, he accorded an important place to Prout's paper (though he did not divulge its author's name) as showing that the volumetric and atomic ways of looking at chemistry need not be in conflict.[11] He also drew attention to Ampère's paper containing his independent assertion of Avogadro's Hypothesis; of which Thomson remarked uncomprehendingly 'this is precisely the law of Mariotte'. In 1818 he published a table of atomic weights as the beginning of his experimental testing of Prout's view; or as he put it, 'to prevent our indulging in theory before we have obtained the requisite data'.[12] This warning should perhaps put us on our guard; and we notice how many of the values, on his scale O=1, are whole numbers, although all values are given to three decimal places.

Those that are not whole numbers are curiously rounded fractions; .125, .375, .625; in fact, eighths, or
multiples of the value for hydrogen, 0.125. To produce such very happily concordant results, Thomson
must surely have been 'indulging in theory'; even
schoolboys cooking experiments soon learn that for
experimental results to be credible they must diverge
a little from the exact values predicted theoretically.
Many of his contemporaries rounded off their atomic
weight values, among them for example Davy; but they
did this for simplicity in calculations, and to
indicate that the results were only moderately
accurate.

Thomson, on the other hand, with his values
quoted to three decimal places, laid himself open to
attack. There is no reason to suppose that he was
indulging in fraud, like the 'discoverers' of the
Piltdown man; rather, in a period when the estimating
of accuracy was not developed, he was choosing results
to bring the average value to where he knew that it
ought to be. In 1818 Thomson moved to a chair of
Chemistry at Glasgow, where he became increasingly
embroiled in academic politics;[13] he was helped by
others in the editing of the Annals of Philosophy,
and in 1821 transferred the editorship to Richard
Phillips. At Glasgow Thomson set his students to
perform research projects, thus anticipating Liebig's
methods at Giessen; but whereas Liebig was to let his
students follow their own lines of research, Thomson's
were made to do his research for him and it seems
that much of what he later published as his own was
the work of his students.

In 1820 there appeared in Annals of Philosophy
detailed analyses by Berzelius, and also some remarks

164

by Thomson.[14]   He wished that Prout would keep up
his chemical work, for he had the experimental skill
and mathematical competence which modern chemistry
demanded; but in the event Prout turned to physiology,
discovering hydrochloric acid in the stomach.
Thomson added that chemists hitherto had performed
chemical analyses without general principles to guide
them:

> hence the very great diversity
> apparent in the numbers on which
> they have pitched, and the com-
> plicated nature of the numbers
> themselves;whereas it will be easy
> to show, when we have established
> the true specific gravities of the
> gases, that the numbers denoting
> the relative weights of the atoms
> of those elements that enter into
> the constitution of the gases are
> of the most simple kind, and that
> they are possessed of properties,
> which, if duly studied, would soon
> lead   us a great way into the
> knowledge of the constitution of
> the vegetable and animal kingdoms.

We are reminded of Beddoes' hope that clearing up the
problem of the elements would advance medical science;
but Thomson's remarks are in the direct line of
development of organic chemistry.

   Apart from the remarks about simple relations,
Thomson did not enlarge upon what he meant by the
general principles which should guide the analyst.
But it must have been clear to his readers what he
had in mind; and he made it sufficiently apparent in
remarking of Berzelius that he 'seems never to have
thought of any relations subsisting between the atomic
weights of different bodies, but to have trusted
entirely to experiment'. Berzelius was no arid
positivist, but a great theorist who succeeded in

combining the dynamical chemical views of Berthollet
and the electrochemists with the atomic theory of
Dalton into a synthesis, dualism, which dominated
chemistry for a generation and brought order into
wide regions of it. To us it is axiomatic that
atomic weight determinations should be based entirely
on experiment, and we have the theory of errors to
help us fix upon the mean value from a set of observations. But just as astronomers could not agree on
which of the many observations of the Transit of
Venus were the best, and should be used in the calculations, so it was not apparent which results the
chemist should use to calculate his atomic weights.
Thomson might after all have been right; individual
experiments do not precisely confirm Boyle's Law or
Ohm's Law, and yet we may believe that the law is
closer to the real state of affairs than the observations, all of which are distorted by experimental
errors. Thomson and Prout might have been like Snell
and Descartes, who discovered the simple sine law of
refraction; while Berzelius might have been like
Ptolemy and Kepler, who drew up tables of angles but
failed to generalise them and reveal the simple law.

In 1821 Thomson set about explaining his
position vis-à-vis Prout's hypothesis.[15] The analyses
of Wollaston and Berzelius were never rigidly exact
though they were usually good approximations.
Berzelius' methods lacked precision, for although his
results were consistent he had no means of checking
himself, no absolute or external criterion of accuracy.
For Thomson, who knew the result in advance like a
schoolboy doing an exercise, things were much easier;
he knew that the atomic weights would be multiples of
that of hydrogen, and the best experimental results
were accordingly those which led to such values.

Thomson used a method of precipitation; for example, he mixed solutions of potassium sulphate and lead nitrate and then determined whether the solution contained 'any sulphuric acid or any [lead] oxide' — he adhered to the ordinary dualistic theory of salts. If it did, then one component was in excess, and the end-point was found by trial and error; Thomson described the process as tedious but foolproof. He did not take all the precautions which Berzelius found necessary, for he was not interested to the same extent in getting concordant results; he wanted those which fitted Prout's hypothesis, his absolute criterion of accuracy.

Thus he wrote:

> I am thoroughly persuaded that as
> we proceed in our investigations,
> the simplicity of the atomic theory
> will become more and more apparent.
> The complex numbers of Berzelius
> will all disappear; and the
> application of mathematical reason-
> ing will by and by enable us to
> advance with unexpected rapidity
> in the chemical investigations of
> the animal kingdoms.[16]

Berzelius was described as labouring under a delusion; and later in the same volume of the journal, Thomson made a frontal attack upon him. Berzelius gave his atomic weights to five decimal places; Thomson wrote that

> the long train of decimals attend-
> ing almost every one of these
> numbers is sufficient of itself to
> render their accuracy suspicious.
> Nature is not wont to indulge in
> such complicated relations. Indeed
> the very simple numbers, which we
> have already obtained for the atomic
> weights of the 14 bodies formerly
> examined, lead almost irresistably

167

> to the conclusion that the same
> simplicity will be found to per-
> vade the weights of all the
> simple bodies in nature.

And indeed he published new whole-number values for copper, zinc, iron, manganese, nickel, and cobalt.

It would certainly have been possible to rebuke some chemists of this era for publishing results which appeared more accurate than they could in fact have been.[17] Lavoisier for example had sometimes given values of a higher order of accuracy than some steps in his experiments. But to accuse Berzelius of this kind of thing was dangerous, and Thomson seems to have been simply less-dextrous in his manipulation than Berzelius as well as carried away by Prout. One is reminded of Dalton, who rebuked the exact Gay-Lussac for inaccuracy when his results appeared to conflict with the atomic theory. What the episode is beginning to show is that the falsification or con-firmation of an hypothesis is not as easy a matter as it is sometimes made out to be. Prout's hypothesis seems eminently falsifiable, in contrast to the 'metaphysical' analogical arguments of Davy and others; all that is necessary is to determine a few, and in the end all, atomic weights and to see whether they are multiples of that of hydrogen or not. And yet on this factual and seemingly-simple issue the most eminent analytical chemists of Europe could argue for more than a decade, attaining a high degree of acerbity in the process; and the controversy was to rumble on through much of the nineteenth century, and only to wither away in the twentieth when the nuclear atom and the doctrine of isotopes put the whole question in a different light.

Thomson in these papers nowhere comes out

explicitly in favour of a theory of the complexity of the chemical elements. Nature would apparently be simple and economical enough for his taste if she produced all the phenomena from a number of simple bodies, provided that their atomic weights were multiples of that of hydrogen. But in Thomson's Annals of Philosophy for 1821 there did appear the pioneer paper on the kinetic theory of gases by John Herapath in which he did advocate but one kind of atom;[18] an absolutely hard one:

> One of the sublimest ideas of the
> ancients was, that there is but
> one kind of matter, from the
> different sizes, figures, and
> arrangements of whose primitive
> particles, arises all that
> beautiful variety of colour,
> hardness and softness, solidity
> and fluidity, opacity and trans-
> parency, &c. which is observed
> in the productions of nature. Our
> ... postulata do not...require that
> there should be but one kind of
> matter; there may be several kinds.
> But since it seems possible to
> account for all the phenomena on
> the supposition of one kind only,
> and since nature is always disposed
> to employ the simplest machinery,
> probability is strongly in favour
> of the ancient idea.

Herepath's theory can perhaps be seen as an attempt to give a quantitative form to the doctrine of the unity of matter; but his mechanics of hard body collisions was unsatisfactory. To the chemist, the differences between substances cannot but obtrude themselves; he needs a plausibly detailed account of the genesis of chemical properties, where the mathematical physicist Herapath puts his '&c'.

In 1825 Thomson brought together his work on Prout's hypothesis in a book grandiloquently titled

An Attempt to Establish the First Principles of
Chemistry by Experiment.[19] The book contained 747
analyses performed by Thomson or by his pupils under
his direct supervision. He wrote that the difficulties were less in the analyses than in the obtaining
of pure starting materials; and that it was here that
Berzelius — to whom credit was nonetheless due for
establishing the atomic theory — went astray, for he
had used allegedly-pure metals as his starting point
instead of really-pure neutral salts. To Prout was
given the highest praise; his paper 'displayed a
sagacity that has seldom been exceeded in chemical
investigations, and shows clearly that the author, if
he chose, might rise to the highest eminence as a
chemical philosopher'. In his own view, Thomson's
investigations confirmed and indeed 'fully establish
the truth of Prout's sagacious conjecture'. Thomson
did not follow Prout in believing that hydrogen was
the only true element:

> It is not likely that any of the
> ultimate elements has ever come
> under our inspection. All our
> simple bodies are most probably
> compounds. It is possible that
> the ultimate elements of bodies
> may be very few — it is even con-
> ceivable that they may be reduced
> to two; but in what way all the
> variety of bodies with which we
> are acquainted, could be produced
> from one single kind of elementary
> body or atom, I cannot, for my own
> part, form any conception.

Thomson's atoms were chemical rather than physical;
they were 'the ultimate particles of which any body is
composed, without considering whether the farther
division of these particles be possible or not'. For
Thomson, the term 'atom' can therefore be properly

170

applied to compounds as well as atoms; and it is
'squeamish' to avoid it and use such circumlocutions
as 'equivalent' or 'proportion'. Berzelius came in
for attack once more; his system was so complicated
and intricate that it could hardly be a true representation of nature.

The book received a mixed press.[20] In the Annals
of Philosophy, the reviewer disagreed with some parts
of it but his conclusion was favourable; the work
should form a part of every chemical library, and
would become a standard work on atomic weights and
the means of arriving at them by experiment. In
America, Silliman's Journal also gave a favourable
review, part of which was reprinted in Annals of
Philosophy; to us it sounds ironical, but it was meant
in all seriousness: 'The consummate skill discovered
in devising and executing the experiments, and the
surprising coincidence of the results of analysis
with the deductions of theory, excite our astonishment,
and prove beyond a question that chemistry, if not
founded on intuitive, is built on demonstrative truth'.

In the Quarterly Journal of Science,[21] edited by
Brande at the Royal Institution, there appeared a
review of the work by Andrew Ure, who is best known
as an apologist for the factory system. Ure had
earlier slated the sixth edition of Thomson's famous
System of Chemistry; he also came from Glasgow, where
he was a Professor at the technological Andersonian
Institution, and he and Thomson seem readily to have
descended to abuse and vilification. Ure attacked
the First Principles on two grounds; he suggested that
a number of the results were cooked or trimmed, or
perhaps wholly imaginary. But this was not the
criticism which he regarded as most serious. The

171

body and soul of chemistry do not reside, he averred, in a knowledge of the combining weights of substances as Thomson supposed. For Ure, the characteristic of chemical genius is to reveal new elementary bodies, new qualities, new relations between bodies, and to arrange under general laws the phenomena of corpuscular action. It is curious to find in a technologist this view of chemistry as a science like natural history; for Ure, quantitative research must always be subordinate to qualitative, and Thomson's kind of enquiry hardly merited the name of an intellectual operation. What was needed was research into the powers that modify matter, in the tradition of Davy, Berthollet, Wollaston, and Gay-Lussac; there was in Thomson's book, despite its bold title, no Herculean achievement like that with which Davy astonished the world in his Elements. Even granted his lowly view, Thomson was imprudent, in Ure's opinion, in using novel experimental techniques to support a controversial proposition; the methods established by Davy and Berzelius were both generally accepted and superior. Finally, and reasonably, Ure took Thomson to task for not using the scale of weights based on H=1, and thus reaping the full advantage of his hypothesis by having whole-number values throughout.

Thomson replied at length, justifying analytical chemistry in general and his own work in particular; and explaining the use of the scale based upon O=1 with reference to the very large number of different bodies with which oxygen will combine.[22] Ure's criticisms of Thomson seem harsh, for Thomson's interests did go beyond the determination of atomic weights; and while his was not an original or very powerful mind, he was a good teacher and the compiler of an excellent and up-to-date textbook. Perhaps he

fell between two stools, in being neither simply a
plodding analyst nor an original thinker. But beneath the personal dislike which Thomson and Ure
felt for one another, there was a genuine disagree-
ment about the way chemistry ought to go. The science
was becoming increasingly technical; matters of
manipulation, and accurate determinations of quantities were coming to the fore. It was becoming less
easy to take an interest in general questions, and
the large audiences which about 1800 had flocked to
hear about advances in chemistry were now to be found
listening to expositions of geology, a science in
which the amateur still had a role to play.

Despite Ure's innuendo, the First Principles
seems to have been initially a success. Thus in 1827
Prout, reading to the Royal Society a paper on organic
analysis, said of his hypothesis: 'though ... drawn
up and published in a very hasty manner, ...[it]
attracted some notice; and the views therein advanced
gradually gained ground, and at present appear to be
generally admitted in this country'.[23] In support of
this claim he mentioned Thomson, Henry, and Brande;
but Thomson must have been far and away the most important. In the same volume of the Philosophical
Transactions Thomson published a paper on the compounds of chromium, in which he referred to Prout's
Hypothesis as having been confirmed in the First
Principles.

The storm broke in the following year, for a
review by Berzelius of Thomson's book was published
in the Philosophical Magazine, which had in 1827 incorporated the Annals of Philosophy.[24] The editor,
referring to Thomson's book, remarked

that the correspondence between his

173

hypothesis and the result of his
experiments is startingly precise.
As the accuracy of his results,
which, if true, are very important,
can be duly estimated only by an
analyst of extensive experience, we
looked forward with impatience to
hear the opinion of Berzelius. His
opinion has at length reached us;
and as it is expressed in language
extremely strong and extremely
unusual, we think it necessary to
employ his own words. We have
accordingly translated a few
passages from the Yahres-Bericht
for 1827 (Woehler's Translation).

The passage, and Thomson's reply, are valuable to the
teacher who wants to demonstrate that scientists are
neither inarticulate and dessicated calculating
machines, nor creatures so cool and dispassionate as
to rejoice when their hypotheses are falsified.
Berzelius said that most of the experiments had been
made at the writing-desk; that the methods described
were clearly bound to lead to erroneous results,
'errors of such a nature as it is difficult to commit,
and which appear to prove that the results were invented'. Thomson had neglected well-known precautions;
his work was 'quackery', and 'the greatest civility
which his contemporaries can show its author, is to
forget that it was ever published'.

Thomson tried to reply; his language was equally
violent, but in any appeal to experimental technique
Berzelius was clearly the highest authority.[25]
Thomson begins to appear a pathetic figure, and his
brash remarks about Berzelius' excessive devotion to
accuracy, made over the previous ten years, now
acquire a certain irony. Berzelius must have been
stung sufficiently by such remarks, and by what he
felt to be neglect of his merits by British chemists,

174

to write such a cutting review; and to have felt
that the science was threatened by such a priori
methods, and such analyses backed by allegedly
absolute criteria of accuracy.

The only solution to such a clash between two
prominent analysts was for someone else to perform
the analyses; and this task was taken in hand by
Edward Turner, who became Professor of Chemistry at
University College, London.[26] He was well-suited to
it, being somewhat plodding but thorough; his results
began to be published in 1829. He declared that 'the
era of brilliant discovery in chemistry seems to have
terminated for the present. The time is arrived for
reviewing our stock of information, and submitting
the principle facts and fundamental doctrines of the
science to the severest scrutiny'. In fact, we should
try to find the exact ratios of elements in compounds.
He commented on the acrimony with which Ure and
Berzelius had assailed Thomson, and avoided it himself. Barium chloride was crucial in Thomson's system,
for he had used it to find the equivalent weight of
sulphuric acid, and of thirteen metals and their
protoxides; and according to Berzelius, Thomson's
analysis of barium chloride (and hence all these
others too) was inaccurate. Turner confirmed
Berzelius' view, noting that Thomson had neglected
standard precautions; in particular, he had failed to
appreciate that when barium sulphate is precipitated
it always carries down some potassium sulphate, so
that the precipitate weighs more than it should, or
than it does when sulphuric acid is used instead of
potassium sulphate solution. This error is not completely avoided even by careful washing of the
precipitate. Turner also found that despite Thomson's
claims, he had used impure starting materials in his

175

analyses.

In 1832 Turner read a paper containing further analyses to the British Association at Oxford; this appeared in the Philosophical Magazine, and an extended version was published in the Philosophical Transactions in the following year.[27] Turner noted that Thomson had in the seventh edition of his System changed the atomic weight of barium from 70 to 68, thus bringing it into line with Berzelius' value; but that, as a result of this, 'many of the experiments described in the First Principles of Chemistry are now at irreconcileable variance with each other, and if relied upon at all, subvert the conclusions which they once appeared to establish'. The equivalent weights of lead and of chlorine, for example, become 103.6 and 35.33-35.42. While whole numbers might be adequate for medical students, 'Dr.Prout's hypothesis, as advocated by Dr.Thomson, — that all atomic weights are simple multiples of that of hydrogen, — can no longer be maintained'. Turner did not despair of finding that there was some relation, but simply affirmed

> that the experiments by which it
> has been attempted to prove the
> truth of this hypothesis are in-
> accurate ... I deny not that some
> simple relation exists among
> atomic weights, and that their
> ratios may possibly be expressed
> by some simple series of numbers;
> but at present no one has assigned
> any physical cause for the exist-
> ence of such a relation; no such
> relation has hitherto been discover-
> ed; nor, as appears to me, has
> analytical chemistry attained that
> degree of perfection which can
> justify anyone in finally asserting
> or denying its existence.

Turner was right; Prout's hypothesis, with
various qualifications, survived apparent falsification, but Thomson as a chemist was done for. As
Babbage remarked in his famous discussion 'On the
Frauds of Observers', 'the character of an observer,
as of a woman, if doubted is destroyed'.[28] Thomson
became increasingly immersed in agitation for reform
of Glasgow University, where some of the Chairs, including that which he occupied, had a lower status
than others and their occupants little say in academic
affairs.

Meanwhile Prout, who had never interested himself
in his hypothesis to the extent to which Thomson had,
backed away from his earlier views, and although the
sub-molecules of his Bridgewater Treatise might be
held to be a prime matter more basic than hydrogen
there is no suggestion that their existence could be
proved from atomic weight determinations.[29] In 1831
Charles Daubeny, Professor of Chemistry and Botany at
Oxford, published what became a standard little work
on the atomic theory. He was tempted by Prout's
hypothesis, but made the qualifications — necessary
since Turner and Berzelius had begun to undermine
Thomson's results — that the basic substance need not
be hydrogen:

> If we could ascertain that there
> were any two or three undecompounded
> substances, to whose atomic weight
> those of all other bodies in nature
> bore this relation, we might with
> some degree of plausibility conject-
> ure, that these were the elements
> out of which the rest were formed;
> but there is no necessity for assum-
> ing that one of them would prove to
> be hydrogen, or even any other of
> the bodies which have yet come under
> our cognisance.

Despite this caveat, he used the scale based upon H=1 in case all atomic numbers were integers on it.

Daubeny corresponded with Prout and showed him the sheets of the book; and Prout's comments on them were added as an appendix.[30] He allowed that his 'speculations' were indeed that all elements might be composed of hydrogen, but that hid 'opinion' was simply that there might be relations between atomic weights, and that there might be bodies still lower in the scale than hydrogen of which all bodies might be multiples. This move makes the hypothesis unverifiable (for the postulated bodies might weigh one tenth, one hundredth, one thousandth, and so on, that of an atom of hydrogen; smaller bodies being invoked each time analysis becomes more accurate) but was to be reasserted in France later in the century. Prout himself plays little further part in this story; his fame rests upon his haunting hypothesis, but his serious work was not in analytical chemistry.

In 1844 an editor of Silliman's American Journal wrote to Berzelius asking him his opinion of Prout's hypothesis. Berzelius replied that the hypothesis had been founded upon a belief in the inaccuracy of experiments; that with a little trimming here and there, within the limits of experimental error, the atomic weights could all be made into multiples of that of hydrogen. Prout had 'given no explanation of the supposed fact on which his theory is based. Such a result could hardly proceed from any thing but the existence of only a single ponderable element — hydrogen — of atoms of which combined, the compound molecules of all the other so-called elements must consist — each of some definite number'. Berzelius' English is not very clear, but he implies that if this

were so, then alchemists or modern chemists would
surely have been able to produce a genuine transmutation; but none have, and 'the elements have remained unchanged — incommutable'. There was, in short,
neither an adequate empirical basis — and what there
was, recent atomic weight determinations had
destroyed — nor a sound theoretical reason behind
Prout's hypothesis; there may be some relations between the atomic weights of certain related elements,
but no general relationship is visible. Berzelius
did not even suggest that the existence of Prout's
hypothesis had been a great stimulus to accurate
experiment.[31]

Berzelius clearly supposed that he had written
the last word on an unedifying but perhaps instructive episode; but in 1853 the chemist J.H.Gladstone
published, in the Philosophical Magazine, a paper he
had written with the mathematician Augustus de Morgan,
who is remembered chiefly as a logician.[32] They had
established that the distribution of atomic weights
was not random; far more were close to being multiples
of that of hydrogen than would be the case in a chance
distribution. While Prout's hypothesis was not
accurate, there would seem to be some basis for believing in a version of it. And at much the same
time, chemists were taken aback by claims that what
Berzelius had challenged believers in the unity of
matter to do had been accomplished; that transmutations had been observed in the laboratory. We shall
turn to these qualitative chemical arguments for the
unity of matter in the next chapter.

NOTES

1. S.Parkes, Chemical Catechism, 10th edn(1822)p.
220; a review of the 13th edn. appeared in Phil.Mag.,
VI(34d series,1835)214; Darwin enjoyed this work and
that of Henry; F.Darwin, Life and Letters of Charles
Darwin, 3rd edn(1887)I,p.35.

2. W.Henry, The Elements of Experimental Chemistry,
7th edn(1815)I,p.xliii; T.Griffiths, Chemistry of the
Four Ancient Elements(1842)p.14; J.F.Daniell, An
introduction to the study of chemical philosophy(1839)
p.268; R.Kane, Elements of Chemistry,2nd edn(Dublin,
1849)p.2; E.Turner, Elements of Chemistry,3rd edn
(1831)p.6. Contrast M.Somerville, Molecular and Microscopic Science(1869)I,p.3.

3. D.Stewart, Philosophical Essays (Edinburgh,1810)
pp.xiii,xv. See A.Thackray, Matter and Powers
(Cambridge,Mass.,1970)p.248; J.B.Morrell, 'The
University of Edinburgh in the Late Eighteenth Century'
Isis,LXII(1970)158-71. Contrast with Stewart a writer
who in 1828 thought sodium and potassium 'little
better than hypothetical assumptions', their properties being so different from those of proper metals
that 'nothing but the rage of the day for the invention of new metals could have prompted their insertion
in the list', C.Singer,et al., A History of Technology
(Oxford,1954-8)IV,p.118. R.Chenevix, Phil.Trans.,XCIII
(1803)290-320; Remarks upon chemical nomenclature
(1802)p.67, and see note 33.

4. A.Comte, The Positive Philosophy,tr.H.Martineau
(1853)I,pp.16,314,331,289,300,308,310,327; cf on
distribution, G.Wilson, Electricity and the Electric
Telegraph, together with the chemistry of the Stars,
new edn(1859)pp.34-6; on Comte and chemistry, see W.H.
Brock, The Atomic Debates(Leicester,1967)pp.145-52.

5. J.M.Good, The Book of Nature(1826)I,pp.59,70,69;
note his discussion of affinity, gravity, and electricity,pp.105-8. Cf W.Whewell's demonstration that all
matter is heavy; Philosophy of the Inductive Sciences
(1847)II,pp.624-34. In the posthumous American edition
of the Book of Nature(New York,1833) the editor compared Good to Boerhaave and Haller; pp.xv,xvi.

6. Anon[W.Prout],Annals of Philosophy,VI(1815)321-30;
VII(1816)111-3; reprinted in Alembic Club Reprints,XX
(1932), with papers by Stas and Marignac; and in fac-
simile in D.M.Knight,Classical Scientific Papers —
Chemistry, Second Series(1970) with papers by Thomson,
Berzelius, and Turner. See W.H.Brock, 'Studies in the

180

History of Prout's Hypothesis', Annals of Science,
XXV(1969)49-80, 127-37. For the atoms and volumes
controversy, see Alembic Club Reprints,IV(1961);
J.J.Berzelius, Annals of Philosophy,II(1813)443-54;
III(1814)51-2; V(1815)122-31; and Dalton in the same
journal,III(1814)174-80. H.Davy, Elements of
Chemical Philosophy(1812)p.481; B.C.Brodie,Chemical
News,XV(1867)295-305, see below,chapter VIII.

7. W.Prout, Bridgewater Treatise,2nd edn(1834)pp.74-
6. On Avogadro's Hypothesis, see the paper by M.
Fricke in C.Howson(ed.), Method and Appraisal in the
Physical Sciences(Cambridge,1976)pp.277-308.

8. On Thomson, see S.H.Mauskopf, 'Thomson before
Dalton', Annals of Science,XXV(1969)229-42; A.
Thackray, 'The origins of Dalton's chemical atomic
theory',Isis,LVII(1966)35-55, 'The emergence of
Dalton's chemical atomic theory', British Journal for
the History of Science,III(1966)1-23; and extracts
from Dalton, Thomson, and Wollaston in Alembic Club
Reprints,II(1961). On the reception of Dalton's
atomic theory, see D.S.L.Cardwell(ed.), John Dalton
and the Progress of Science(Manchester,1968) —
especially in our context essay 14, by W.H.Brock;
D.M.Knight, Atoms and Elements(1967); and A.L.Smyth,
John Dalton,1766-1844:a bibliography of works by and
about him, (Manchester,1966). J.J.Berzelius, Annals
of Philosophy,III(1814)51-2; T.Thomson, Annals of
Philosophy,II(1813)32-43. J.Morrell, 'The Chemist
Breeders:The Research Schools of Liebig and Thomas
Thomson', Ambix,XIX(1972)1-46.

9. T.Thomson, Annals of Philosophy,II(1813)33;III,
(1814)331; H.Davy, Phil.Trans.,CI(1811)15, footnote,
going over onto subsequent pages; T.S.Wheeler and
J.R.Partington, The Life and Work of William Higgins,
Chemist(Oxford,1960).

10. T.Thomson, Annals of Philosophy,IV(1814)11; III
(1814)XIII; XII(1818)6 and New Series,XI(1826)1ff.,
H.Davy, Works,VII,pp.92-9.

11. T.Thomson, Annals of Philosophy,VII(1816)17;
Prout is identified on p.343, and Ampère's theory on
p.19.

12. T.Thomson, Annals of Philosophy,XII(1818)338,436;
Davy gives whole 'numbers representing the undecomposed bodies' (his circumlocution for atomic weights) in
his Elements of Chemical Philosophy(1812); in his
Elements of Agricultural Chemistry,2nd edn,(1814)p.46
he gives a fractional value for carbon. Dalton's
figures in his New System of Chemistry,part I(Manchester,1808)p.219 are whole numbers; but Wollaston's,

Phil.Trans.,CIV(1814)18-23 are not; Berzelius criticised 'English chemists' use of round numbers in
Annals of Philosophy,II(1813)453-4.

13. See the articles by J.B.Morrell, 'Thomas Thomson:
Professor of Chemistry and University Reformer',
British Journal for the History of Science,IV(1969)
245-65, 'Practical Chemistry in the University of
Edinburgh,1799-1843',Ambix,XVI(1969)66-80.

14. J.J.Berzelius, Annals of Philosophy,XV(1820)89,
276,352; T.Thomson, Annals of Philosophy,XVI(1920)167,
177,247; T.Beddoes, Contributions to Medical and
Physical Knowledge(Bristol,1799)p.230. See also J.J.
Berzelius, Essai sur la Theorie des Proportions
Chimiques(1819),reprint,intr.C.A.Russell(1973); and
H.Wolff, The Transits of Venus(Princeton,1959).

15. T.Thomson, Annals of Philosophy,new series,I(1821)
3.

16. T.Thomson, Annals of Philosophy,new series,I(1821)
15,242.

17. Edward Turner, Phil.Trans.,CXXIII(1833)543 while
confirming Berzelius' results also suggests that he
gives his answers to too many decimal places.
Lavoisier's study of vinous fermentation, for example
(Elements of Chemistry,1790,ch.13) seems to contain
results quoted to a higher degree of accuracy than
was really attained. J.Dalton, Elements of Chemistry,
part II(Manchester,1810)pp.555-9 — also in Alembic
Club Reprints,IV(1961)25-7. Dalton's experimental
work is now more highly regarded than it used to be;
L.Trengove, 'Dalton as an Experimenter', British
Journal for the History of Science,IV(1969)394-8;
D.S.L.Cardwell(ed.), John Dalton and The Progress of
Science(Manchester,1968)essay 10, by K.V.Farrar —
see also essay 2, 'The Qualitative and the Quantitative', by C.N.Hinshelwood. For similar problems in a
different context, see C.D.Hellman, 'The Role of
Measurement in the Downfall of a System', in A.Beer
(ed.), Vistas in Astronomy(Oxford,1968)pp.43-52.

18. J.Herapath, Annals of Philosophy,new series,I(1821)
281,see also his Mathematical Physics(1847). S.G.
Brush, 'The kinetic theory of gases,I, Herapath',
Annals of Science,XXIII(1957)188-98, Kinetic Theory
(Oxford,1965-6); G.R.Talbot and A.J.Pacey, 'Some early
kinetic theories of gases:Herapath and his predecessors
British Journal for the History of Science,III(1966)
133-49. For the history of hard-body collisions, see
W.L.Scott, The Conflict between Atomism and Conservat-
ion Theory,1644 to 1860(1970); E.Mendoza, 'Herapath's

Dynamical Theory of Gases', British Journal for the History of Science,VIII(1975)155-65.

19. T.Thomson, An attempt to establish the first principles of Chemistry by experiment(1825)I,pp.xv, 22,24,26,31,35,xviii.

20. Annals of Philosophy,new series,X(1825)138,147; XI(1826)68.

21. A.Ure, The Philosophy of Manufacturers(1835); The Quarterly Journal of Science,XX(1825)113-9; cf H.Davy, Works,VIII,p.287, and contrast J.G.Children's introduction to his translation of L.J.Thenard's Essay on Chemical Analysis(1819)p.viii,and The Quarterly Review,XLV(1831)392, favouring quantitative chemistry.

22. T.Thomson, Annals of Philosophy,new series,XI (1826)1ff. See D.M.Knight, 'Chemistry in palaeontology: the work of James Parkinson',Ambix,XXI(1974)78-85; and D.M.Rudwick, The Meaning of Fossils(London,1972).

23. W.Prout, Phil.Trans.,CXVII(1827)355; T.Thomson, Phil.Trans.,CXVII(1827)132.

24. J.J.Berzelius, Phil.Mag.,IV(2nd series,1828)450-3; for his earlier advocacy of accepted methods, see Annals of Philosophy,V(1815)129.

25. T.Thomson, Phil.Mag.,V(2nd series,1829)217-23. There is some irony in the quarrel in that important paper by Berzelius had appeared in English in Thomson's Annals of Philosophy. On Berzelius' accuracy, see F.Szabadvary, History of Analytical Chemistry,tr.G.Svehla(Oxford,1966)p.149. See J.J. Berzelius, Annals of Philosophy,VI(1815)47 for complaints of neglect by British chemists.

26. E.Turner, Phil.Trans.,CXIX(1829)291-300.

27. E.Turner, Phil.Mag.,I(3rd series,1832)109-12; Phil.Mag.,CXXIII(1833)523,544.

28. C.Babbage, Reflections on the decline of science in England(1830)p.182; on Thomson, see anove note 13.

29. W.Prout, Bridgewater Treatise,2nd edn(1834)pp. 148ff; C.Daubeny, An Introduction to the Atomic Theory (Oxford,1831)pp.39,42.

30. C.Daubeny, An Introduction to the Atomic Theory (Oxford,1831)pp.129-33; Alembic Club Reprints,XX (1932)58 for Marignac on the unity of matter; and see below, chapter VIII.

31. J.J.Berzelius, Silliman's American Journal of Science,XLVIII(1845)369-72; quotations from pp.369

and 370.

32. J.H.Gladstone, Phil.Mag.,V(1853)313-20.

33. M.C.Usselman, 'The Wollaston/Chenevix controversy over the elemental nature of palladium', Annals of Science, forthcoming.

7

Isomerism and Isomorphism

The Irish chemist Michael Donovan in the volume on
Chemistry in Dionysius Lardner's series, The Cabinet
Cyclopedia, referred to Prout's hypothesis, remarking
that most equivalent weights were whole numbers.[1]
This singular circumstance, he remarked a little
nastily, might be accounted for 'by supposing that
the atom of hydrogen is the natural and real unit of
all matter; or, more probably, that fractions have
been removed by the corrections made in the analyses
of bodies, according to a practice now almost invar-
iably adopted and followed by chemists'. We have
seen that he was right, and that accurate analyses
showed that Prout's Hypothesis was in its original
form false. The first volume of the Cabinet Cyclopedia was the Preliminary Discourse of Sir John
Herschel, the son of Sir William, which became a
standard work on philosophy of science and scientific
method, exerting a strong influence on Victorian men
of science. Herschel thought Prout's hypothesis so
important that no labour spent in verifying it would
be wasted; but that

> philosophical chemistry no more aims
> at determining the one essential
> element out of which all matter is
> formed — the one ultimate principle

of the universe — than astronomy
at discovering the origin of the
planetary movements in the
application of a determinate pro-
jectile force in a determinate
direction, or geology at ascending
to the creation of the earth.
There may be such an element.
Some singular relations which have
been pointed out in the atomic
weights of bodies seem to suggest
to minds fond of speculation that
there is; but philosophical
chemistry is content to wait for
some striking fact, which may either
occur unexpectedly or be led to by
the slow progress of enlarged views,
to disclose to us its existence.

The analysis of an element would be an important and
indeed triumphant achievement, and would change the
face of chemistry, but would undo nothing that had
been done and render no existing knowledge useless.

Herschel's remarks were normative rather than
descriptive, and it would not be difficult to find
eminent astronomers and geologists who had not followed his precepts. Chemists too did not follow his advice to pursue analyses and seek quantitative laws.
In the 1820s and 30s the great problem facing
chemists was to unify organic and inorganic chemistry.
The key to the understanding of organic chemistry —
which down to about 1820 was little better than advanced cookery — was the elucidation of the arrangement of atoms or particles in the various compounds.

We saw that Davy, although he inaugurated electr-
chemistry with his Bakerian Lecture of 1806, came
increasingly to see that such properties as acidity
or crystalline form depended not simply upon the
electrical or chemical nature of the component
particles but on their mode of arrangement. The more
systematic Berzelius, by contrast, adhered through

thick and thin to the doctrine that it is the electrical state of the atoms in a compound which chiefly determines its properties. Berzelius was sceptical about the unity of matter, especially after his demolition of Prout's hypothesis; but Faraday carried on the tradition of Davy, disliking the term 'atom' and throughout his working life giving his support to belief in the unity both of force and of matter. Neither Davy nor Faraday seem to have taken much interest in the fate of Prout's hypothesis; they were more impressed by qualitative and analogical arguments.

Ironically it was in Berzelius' laboratory that two crucial concepts for understanding the arrangements of atoms were framed; namely isomerism and isomorphism. These, with the doctrine of Liebig and Dumas that in organic chemistry compound radicals play the role which in the inorganic realm is played by elements, were to provide the analogies required by those who asserted the complexity of the chemical elements.[3] In 1824 Liebig and Wöhler analysed fulminic and cyanic acids respectively, and noticed that the analyses were identical; and at about the same time Faraday discovered butylene, which seemed to have the same composition as ethylene. Berzelius in 1832 generalised and explained these observations, distinguishing the two cases:

> in physical chemistry it was long
> taken as axiomatic that substances
> of similar composition, having the
> same constituents in the same pro-
> portions, necessarily must also have
> the same chemical properties. The
> investigations of Faraday appear to
> indicate that there may be an except-
> ion to this if two similarly composed
> substances differ in that the
> composition of one contains twice as
> many elementary atoms as occur in the

187

other, although the proportions
between the elements remain the
same. ... Recent researches have
now shown that the absolute as
well as the relative numbers of
elementary atoms may be the same,
their combination taking place in
such a dissimilar way that the
properties of equally composed
bodies may be different.

All attempts to prove a difference of composition between cyanic and fulminic acids having failed,
Berzelius went on to name the phenomenon: 'Since it is
necessary for specific ideas to have definite and consequently as far as possible selected terms, I have
proposed to call substances of similar composition and
dissimilar properties isomeric, from the Greek
ἰσομερης (composed of equal parts)'.

Rather earlier than this, Mitscherlich had discovered the phenomenon of isomorphism.[4] The atomic
theory had induced him, he wrote, to investigate
whether different elements united with an equal number
of atoms of other elements would have the same crystalline form; or in other words, whether crystalline form
was the result only of the number of atoms and not of
the nature of the elements. His experiments verified
this hypothesis, to the surprise of adherents to
Berzelius' dualism. Mitscherlich found that the same
substance can sometimes assume two different crystal
forms (this phenomenon he called 'dimorphism') and he
ended his paper with a 'general law for the relation
of crystallography to chemical composition', that 'an
equal number of atoms, if they are bound in the same
way, produce similar crystal forms, and the crystal
form depends not on the nature of the atoms but on
the number and method of combination'. In dimorphism,
the two structures arose from different arrangements

of the same atoms. Isomorphism turned out to be a
less general phenomenon than Mitscherlich's remarks
might imply; but these studies on isomerism and isomorphism marked important extensions of the atomic
theory beyond the explanation of definite proportions
(though it was still far from necessary for the
chemist to believe in atoms), and carried forward ideas
on the arrangement of particles which had begun with
Davy's studies of the diamond, and his investigations
and those of Gay-Lussac on acidity.

In Britain notice was taken of these discoveries,
and an attempt made to use them to cast light on the
problem of the elements, by a pupil of Berzelius,
J.F.W.Johnston, in a Report to the British Association
in 1837.[5] Johnston, a graduate of Glasgow University,
had recently become Reader in Chemistry at the newly-
founded University of Durham and was a fellow of the
Royal Societies of London and Edinburgh. He succeeded
in showing that isomorphism was not a complete guide
to structure; calcium carbonate and potassium nitrate
were isomorphous but on the currently-accepted atomic
or rather 'equivalent' weights did not have similar
formulae, while potassium and ammonium salts were
often isomorphous and yet where the former contained
one atom of potassium the latter had a group of five
atoms (one nitrogen and four hydrogen). Dimorphism
made it clear that crystalline form was neither
specific nor constant; and thus Mitscherlich's discoveries raised more questions than they answered.

Johnston looked for the cause of the phenomena,
and examined the affects of heat, electricity, and
pressure; he found that at a given temperature one
form was preferred by dimorphous substances, and that
electricity and pressure had some influence also.

But these changes were not enough; the cause must be sought in changes of relations among the particles, and he suggested that isomerism might be related to dimorphism, and further that isomerism might explain differences between the elements. The first of these suggestions was taken up by Berzelius[6] in a paper which appeared in English in 1846 in a useful journal entirely filled with translated articles. He remarked that elements appear in alotropic modifications, and suggested that the different forms of carbon contribute to isomerism among their compounds; a version of this doctrine was confirmed by the Braggs in the twentieth century when their X-ray studies revealed the tetrahedral structure of the diamond and the hexagonal rings in graphite. Berzelius listed seventeen elements known in allotropic forms, and suggested that the series of different salts produced by what we call the transition metals might be a consequence of this allotropy. Chromium produced three series of salts, but was only known in two forms; and Berzelius suggested that a third be sought. The paper is chiefly of interest as showing Berzelius in speculative mood, and seems to have born little fruit.

Johnston's other suggestion he developed himself in the paper.[7] If elements like sulphur and carbon appear in allotropic or dimorphous forms, and are thus susceptible of 'internal molecular change', may they not also be susceptible of that deeper change which produces isomerism? The change is deeper because while allotropes differ only in physical properties isomers have different chemical properties too. He used familiar arguments to support this conjecture, an affirmative answer to which would, he believed, be the next great step in chemistry, comparable to the discovery of the alkali metals.

So far, Johnston added, the efforts of chemists had been mostly directed to the decomposition of elements. But it might now be more profitable to leave this traditional approach, and to look instead for transformations. The theory of isomerism, 'rich already in its numerous discoveries, has shown that any number of the received elementary bodies may be made up of the same elements united in the same proportion', for isomers often have very different properties. That no such transformations had been observed did not in Johnston's view justify scepticism, for few isomers had been interconverted either. While hopes for the decomposition of elements depended on the application of galvanism or some even more powerful agent directed by the hand of a Davy or a Faraday, transformations might well be easier and might even be possible though decompositions were not. Three branches of chemistry — analysis and atomic theory, electrochemistry, and crystallography — were all held up by the problem of the elements; would one of them be able to solve it, or would a new branch of science arise?

Johnston does not seem to have followed up his suggestions with actual experiments, but the seeds he scattered did not fall upon wholly stony ground and fruits were not wanting. In turning to workers who speculated and experimented in a manner we cannot but consider faulty, we must remember that the scientist who would make great discoveries must voyage in strange seas of thought. Some of those we shall look at certainly set off ill-prepared for their voyage; and others, though as well-equipped as could be expected, suffered shipwreck. Disasters befall even the greatest of explorers, and their greatness is

shown in surviving them or turning them to advantage;
but that the observers of alleged transmutations in
the nineteenth century were unable to do, and there
is a measure of tragedy as well as of farce in their
histories.

Sir Robert Kane, an editor of the Philosophical
Magazine, put speculations akin to those of Johnston
and Berzelius into his standard textbook.[8] He noted
that there are two forms of chromium, and that the
metal forms two oxides and two series of salts derived
from them, one red and the other green. This 'difference of properties ... might belong to a totally
different substance', and the only reason for not
classifying the two series as coming from different
elements was that one could be changed into the other.
Manganese and copper were other examples of metals
forming different series; and Kane remarked that 'in
order to fully explain the relations which the allotropic condition of bodies bear to their chemical habi-
tudes, we must admit that the same elementary body may
in forming different classes of compounds, abandon
that absolute simplicity and act in a totally different elementary bases'. The manganese which yielded a
basic oxide might be a different substance from that
manganese which with oxygen formed a powerful acid;
and the cupric and cuprous salts differed more than
did cuprous salts and those of silver. Is it not
possible, he asked, that what are called 'the different
degrees of combination' of a metal, represent derivatives of different compound radicals, and that the only
link between them is that from the compounds of both
radicals one can produce the same metal. Further,
different metals were connected by family resemblances
and by relations between their atomic weights; and of

the two manganeses, one had an equivalent weight double that of the other. It seemed therefore possible that metals would be found to be isomers of one another; and that all or most might be compound radicals.

This last view was not new, but it had received considerable impetus from the publication in 1837 of the radical theory by Dumas and Liebig.[9] Inorganic chemistry, they said, had made immense progress in the sixty years since Lavoisier had framed his definition of an element, but this structure did not depend upon the elements being undecomposeable: 'if these elements were later decomposed, nothing would be changed in the architecture of the monument, although its foundations would be more profoundly excavated'. In organic chemistry nearly all compounds were made up of oxygen, hydrogen, carbon, and nitrogen; the units in organic chemistry were not therefore the elements, but certain stable groups of them — the radicals: 'cyanogen, amide, benzoyl, the radicals of ammonia, the alcohols and analogous compounds — these are the true elements on which organic chemistry is founded and not at all the final elements ... which appear only when all trace of organic origin has disappeared'. Their conclusion was splendid in its simplicity: 'In mineral chemistry the radicals are simple; in organic chemistry the radicals are compound; that is all the difference. The laws of combination and of reaction are otherwise the same in these two branches of chemistry'. This paper had been written as part of the attempt to bring into the organic field the order which seemed to prevail in the inorganic; but soon organic chemistry, now the most progressive branch of the subject, began to seem more orderly and simple than inorganic chemistry with all

its different elements. The argument of Dumas and
Liebig was soon turned on its head, and the resembl-
ance between the elements and the compound radicals
was felt to indicate their similarity in structure.
Kane in his textbook remarked of the radical Kacodyle
(in our terms, cacodyl$(CH_3)_2$As-) 'we know its composition, but we do not know the composition of the
simple metallic bases. This is the only philosophical
distinction between the simple and the compound
radicals'.

In 1843 Liebig's <u>Familiar Letters on Chemistry</u>
appeared; this was a charming work, and went through
many editions, swelling rapidly as further letters
were added. In the first edition, Liebig raised in
the letter on isomerism the question of the nature
of the elements; by the third edition of 1851 these
speculations had been dropped. Even in 1843 he was
sceptical,[10] although he asked rhetorically the
question whether, if different arrangements of the
same elements gave rise to different physical and
chemical properties, and similar arrangements of
different elements produced similar physical properties
it was not reasonable to suppose that some at least of
the elements were 'the same matter in a different state
of arrangement?'. But there was a problem; isomers
were usually more unlike in their properties than
related elements are. Liebig kept his feet on the
ground of solid experimental evidence; urging that
until authenticated facts were available, 'no pretensions to conversions deserve notice'.

Such pretensions were current; and were published
in the very respectable <u>Transactions</u> of the Royal
Society of Edinburgh in 1844 by Samuel Brown, M.D.[11]
In 1843 the Chair of Chemistry at Edinburgh had fallen
vacant, and Brown announced his candidature. His paper

had been read in 1841, and he realised that he must substantiate it. In 1843 he accordingly delivered a course of lectures on the atomic theory to the University of Edinburgh, which were said to have attracted an audience 'perhaps the most intellectual ... ever addressed in Edinburgh', which included Davy's brother John, and Sir William Hamilton. Brown was a friend of Robert Chambers, the author of the notorious evolutionary work, Vestiges of the Natural History of Creation.

Brown began by seeking for a proper philosophical definition of chemistry, and a proper scheme for the classification of the chemical elements.[12] He noted that the definition of 'element' was purely contingent, and declared that Lavoisier 'descried the momentous indication that the deeplier the chemical analyst penetrates from the surface of nature, he converges the more until he reach a comparatively small list of elements, perhaps to be lessened and lessened till only two should remain intact...'. What chemistry needed, to make a true 'mixed science' out of it, was an hypothetical premise like the atomic theory, which could be used as a 'fiction of methodology' following the methods of astronomy. The atomic theory thus used would make a deductive science of chemistry; and Brown approved of Exley and Mossotti who had applied mathematics to atomism after the manner of Newton and Boscovich. But if the atomic theory were to be taken seriously and used as anything more than an hypothesis to achieve coherence in chemistry, the results would be tragic. Atomism cannot be a vera causa, for we are unfamiliar with indivisibility; it must therefore be used as a helpful fiction. One must avoid the 'vulgarity of conception' of Democritus and Dalton, while

195

allowing the use of point centres of force in mathematical reasonings.

Indivisibility being inconceivable, and atoms fictions, Brown could hope with Faraday that a new analytical power would decompose the so-called elements and urge that this should be sought diligently 'unless the mind be content to think that, in a mechanism where all is simplicity of design and superstructure, there are at least fifty-five, and likely far more, material bases'.[13] He now came to his experiment in which carbon had been transformed into silicon. He obtained the substance 'paracyanogen' by throwing lead cyanide onto a heated plate; it was a very stable solid isomer of cyanogen. He accounted for its stability by supposing that its formula was Cy_2; that is, that there was no difference between its components to be polarised. It must be composed of two particles of cyanogen revolving about each other on their innermost spheres of repulsion — nearer to one another, that is, than in solid cyanogen. Cyanogen was therefore behaving as an element — its similarity to chlorine had long been noted — and if sulphur were composed, after the manner of paracyanogen, of two oxygen particles revolving close to one another, it might similarly be impossible to decompose it, but possible to synthesise it. Indeed, 'the various elements may all be isomeric forms of one element'. Transmutations, on this view, would only be possible up the series, towards elements of higher atomic weight

It will not surprise us that Brown's heroes were Davy and Prout.[14]

> Davy, the most triumphant experimenter
> that ever lived, constructed a hypo-
> thesis for the purpose of investigat-
> ion, to the effect that the elements

are compounds of hydrogen and an
unknown base in different proportions, and even determined the
proportions of these in the
different elements conjecturally,
as well as performed many
experiments in the attempt to
realise his bold guess.

These remarks refer to Davy's researches of 1808 to
1810, after his isolation of the alkali metals and
before his realisation of the elementary nature of
chlorine. Davy had found hydrogen to be emitted from
various substances such as sulphur; but this later
turned out to be absorbed and not combined hydrogen.

According to Brown, Davy had given up these
speculations when he accepted the law of definite
proportions; and similarly Prout's Hypothesis had been
overthrown by the exact analyses of Turner. But every
young chemist 'of any ardour of hope, and at the same
time prone to divination,... flatters himself he can
reduce the scale to two, and that quite independently
of the speculations of others'. Brown concluded this
part of the lecture with a purple passage:

The great primitive truths of which
the higher theory of nature is
eventually to be constructed, first
appear as mirages among the purpled
morning clouds, casting down their
shadows during a grey afterdawn,
and hovering about the long unsettled
horizon of human thought, often in
most questionable guise; but noon
reveals the coming form distinctly on
the verge, and the predestined craft
shall sail up and over until, ere the
evening of man's day of work, it
reach the shore, awaited with the
timidity of hope, and fraught with
the wealth of invisible climes.

It is not quite clear what time of day it was in 1843.
Brown, like the much greater physicist Thomas Young,

accused Davy of excessive rhetoric, but clearly this was a matter of the pot calling the kettle black.

Brown went on to describe his experiments with his paracyanogen.[15] Though it was stable, it could be decomposed by heat; at first cyanogen was given off, and then nitrogen; and a residue of pure silicon was left. Metallic vessels proved themselves liable to corrosion, and so slow heating in a crucible in a sand bath (twenty days at 800-900°F) was employed. This yield in this novel alchemy was apparently good; a sooty powder was obtained, which was analysed by fusing it with potassium carbonate and separating it as silica. In another experiment, rhodium appeared to have been synthesised. The transformation of carbon to silicon had been repeated 100 times, and Brown awaited with confidence its repetition by others

Brown's conclusions were that chemistry was the science of atoms as existing in objects occupying space, and as agents in phenomena taking place in time. It has therefore a statical and a dynamical, or phenomenal, branch. The elements are all compounds made up of equal and simple atoms and thus perhaps all derived from one basic atom; they form the links of a chain which is 'probably a network wrought into the form of a cone', and fall into isomeric groups of related bodies, like fluorine, chlorine, bromine, and iodine. Such a theory could explain why the elements cannot be decomposed, and — as Daltonian atomism could not — why they are polarised in an electric field. In discussing chemical taxonomy later, we shall come across other examples of three-dimensional arrangements of the elements.

His obituarist tells us that Brown decided to stand for the chair of chemistry at Edinburgh because of the independence, authority, and laboratory

facilities that success would bring. Meanwhile one
chemist had taken notice of his results; a paper by
C.J.Knox on the compound nature of nitrogen had been
read by Kane to the Royal Irish Academy in Dublin.[16]
This also arose out of Davy's experiments; in one of
which he had heated 'ammonia nitruret of potassium' in
an iron tube and found that more hydrogen and less
nitrogen came off than he had expected; when he used
a platinum tube, ammonia was emitted and an alloy of
potassium and platinum formed. Knox believed that in
the first case the nitrogen must have been decomposed
by the metals. He made the substance (potassium
amide) by passing ammonia over potassium in a hot iron
tube, verifying that ammonia and iron together did not
yield silica, and that his potassium contained none;
on heating the amide in an iron crucible, silica was
nevertheless produced, 20 grains each of amide and
iron yielding 1.550 grains of silica after half an
hour's heating. Nitrogen was therefore a compound of
silicon and hydrogen.

This result seemed to conflict with Brown's
results, which to Knox were 'most unreasonable, and
contrary to all chemical analogy; whilst the suppos-
ition of the carbon having reduced the nitrogen is not
only simple but an unavoidable conclusion to arrive at,
if nitrogen be a compound substance'. According to
Knox, then, the equation for Brown's process is:
$N \xrightarrow{C} Si + H$; while for Brown himself it was: $4C \rightarrow Si$.
Knox's mechanism was that of a decomposition, and
Brown's that of a synthesis; the latter had an elaborate theory of spheres of repulsion but could not
explain both sets of results, for Knox had produced
silicon in the absence of carbon.

Little notice was taken of Knox's paper, which

was published in 1844; but when Brown stood for the Chair at Edinburgh Kane offered to act as referee for his procedures. Brown therefore went to Dublin but the experiments failed to work in front of Kane.[17] Meanwhile George Wilson, a friend and fellow-student of Brown's, attempted to repeat them in Edinburgh. Both Kane and Wilson were favourably-disposed towards Brown; indeed when Brown announced his candidature Wilson had written to the Lord Provost declaring his confident expectation that the experiments could be repeated.

Wilson himself later became Professor of Technology at Edinburgh; he is now best known for his collection of essays, Religio Chemici, which included an important study of Dalton, and for his biography of Cavendish which drew William Ramsay's attention to the experiments on sparking air and thus played a role in the discovery of argon.[18] He believed, as he wrote in 1859, that the elements 'do not present that character of completeness and unity, which we should expect in the raw material of a whole universe'. Some elements seemed to stand alone while others fitted into a chain with small gradations of propert-ies; congeners of the former class might be found. He remarked that it had 'been surmised that these so-called elements are only modifications of a primordial form of matter, brought about under certain conditions' in the 1840s he had been more sympathetic to this view, and sceptical of Daltonian atomism.

Wilson's collaborator in his experiments was J.Crombie Brown;[19] their first difficulty was that the atomic weight of carbon was 6 and that of silicon 22.22; so that Samuel Brown's belief that four carbon atoms closely packed yielded one of silicon was not true in quantitative terms. But more serious was the fact that his experiments would not work. After six weeks of attempts they still found it impossible to prepare pure lead cyanide by the method given by Brown, and such residues as they did get did not behave as Brown claimed that they should. None of his processes yielded silicon in good yield, though sometimes gritty residues which probably were silica were found; and Wilson could only conclude that Brown's results could not be substantiated. Liebig remarked that Brown's descriptions of his processes had anyway revealed his incompetence, and that 'his rhodium is iron, and his silicon an impure incombustible coal'. Brown withdrew from the contest for the chair; but, perhaps because of the impression made by his lectures, he nevertheless came second, behind Gregory, in the election. Edinburgh University thus narrowly avoided having an extraordinary Professor of Chemistry; and we are told that 'from that time, as the chemist, he was unheard of', but that he never gave up hope of realising a transmutation, working on despite ill-health.

Wilson's reluctant demolition of Brown's pretensions concluded this episode; nobody seems to have felt it necessary to refute those of Knox. But the arguments for the elements being isomers were not thereby invalidated; they were merely shown not to be easily verifiable. Two other workers apparently of some distinction entered the field, the agriculturists David Low and Robert Rigg. Low was Professor of Agriculture at Edinburgh, and in a book devoid of descriptions of experiments sought to demonstrate that the elements must be complex; while Rigg, a Fellow of the Royal Society, produced experimental evidence for the old view that organisms can transmute, and in particular that carbon is synthesised by plants and decomposed in putrefaction.

Rigg's book, Experimental Researches, shewing carbon to be a compound body made by plants, 1844, was a criticism of Liebig's theory of the carbon cycle, and also of the theory of photosynthesis.[20] His starting point was the familiar observation that plants on good soil grow better than those on poor; and yet if they derived their carbon by photosynthesis the nature of the earth in which they are planted should have little effect. If poor soil be manured so as to bring its carbon content up to that of good, it is found that this carbon is used up and the land requires fresh manuring each year. Good soil, on the other hand, requires no manure at all, and maintains its high carbon content indefinitely. On the received view, therefore, plants on good land derive all their carbon from the atmosphere by photosynthesis, while those on poor land take as little as 17 per cent of their carbon from this source. It was possible to cast doubt on laboratory experiments which seemed to confirm photosynthesis;[21] and Rigg's own researches appeared to show that 'plants during the healthy performance of their functions in the ordinary course of vegetation, impart to the surrounding atmosphere more carbonic acid than they take from it'. Other experimenters had been misled either by using damaged or unhealthy plants, or by odd and artificial experiments — such as growing plants in atmospheres containing excess carbon dioxide — which could prove nothing about growth under normal circumstances. The presence of carbon dioxide in the atmosphere was undoubtedly necessary to plant growth; but it did not supply carbon to plants.

If plants did not absorb carbon dioxide but on the contrary generated it, like animals and volcanoes, then there was a new problem: how was it that the concentration of this gas in the atmosphere remained approximately constant? It might be, Rigg thought, that thunderstorms removed some of the excess; but the chief process consuming carbon dioxide was putrefaction, in which carbon undoubtedly disappeared and water was generated. Organic materials containing a high proportion of carbon, like heart of oak (43 per cent) therefore rotted slower than those with less, like leaves (9 per cent). If only, wrote Rigg,

> chemists would make experiments
> more resembling the processes of
> nature, and take into account
> practical results, I think they
> would find in the various facts
> and deductions, enough to convince
> them, that all the carbonic acid
> which disappears from the atmosphere,
> does not enter into the constitution
> of the plants which grow upon the
> earth, and that a large proportion
> of it disappears by some other and
> unknown pricess; and that plants,
> by a process equally unknown, make
> carbon, and furnish to the atmosphere
> fresh carbonic acid in compensation
> for that which disappears from it.

Rigg's final conclusion was that hydrogen was the only ultimate element found in organised matter; and he had even wondered whether hydrogen were a compound.

Rigg performed numerous experiments, and referred to the work of plant physiologists and of agriculturists in favour of his conclusions. But a vague theory, even if experimentally-supported, which laid such stress on the power of living organisms to do what chemists could not, ran counter to the whole tenour of nineteenth-century chemistry. Rigg's work illustrates the difficulty of doing clear and decisive experiments in this field; and the success of Liebig's biochemistry made such speculations seem

203

utterly obsolete, the end of a cul-de-sac the opening of which had been the willow-tree experiments of Nicholas of Cusa and of Van Helmont. In the classic work of Schleiden, <u>Principles of Scientific Botany</u>,[22] in which botany was handled 'as an inductive science', such speculations as Rigg's were derided:

> Of the origin of these substances
> [metals] in plants, and more
> particularly with regard to the
> question as to whether plants take
> them up from the earth, or form
> them by a peculiar process of
> vegetation ... there is but one
> opinion amongst chemists and
> physiologists, and that is, that
> no elementary body can be present
> in plants that has not been taken
> up from without the plant.

The opposite view survived only as 'a curiosity', though it had received support from eminent men and institutions. Thus by 1850 the old doctrine that organisms transmute seems to have died; but we shall notice a preoccupation with carbon among later devotees of the unity of matter. Although Low was an agriculturist, his arguments were all chemical;[23] his book, <u>An Enquiry into the Nature of the Simple Bodies of Chemistry</u>, went through three editions, growing in size each time. The transformation of carbon into silicon was illustrated, in his view, by the petri-faction of wood, though he did not believe that this reaction would take place in the laboratory. He also noted that statues and buildings stand for ages, but when once they are broken and plants take root they disintegrate and seem to be transformed into carbon-aceous matter. His most serious arguments were less ingenious than these; they were based upon his con-viction that there could not be more than two real elements. Berzelius and a reviewer in the

<u>Philosophical Magazine</u> criticised his book as purely
hypothetical; Berzelius' view was that the problem
of the complexity of the elements should be faced
only when analysis had given some evidence of it.
Low's reply was directed <u>ad hominem</u>; he referred to
Berzelius' own speculations on the nature of nitrogen,
and Lavoisier's on the earths.

Low spoke of the 'want of simplicity and harmony'
in the division chemists imposed between elements,
regarded as simple, and compounds. Lavoisier's
definition established no gulf in physical and
chemical properties between these classes; and it
should therefore be replaced by the maxim 'a body is
to be regarded as compound, when we are not able to
prove it simple'. This move calls to mind the famous
'pourquoi non' argument of Fontenelle, who instead of
arguing directly for the existence of rational beings
on other planets put onto his opponents the onus of
proving that there were none. Low felt that the
analogies were all on his side; George Wilson wrote
that while he sympathised with Low's speculative
spirit, and differed from most chemists in believing
that such a view would be confirmed in time in the
laboratory, he was prepared to accept the ordinary
definition of an element, as had Davy and Berzelius
both of whom had speculated and experimented freely
on the decomposition of elements. Low felt that
Davy's position was open to disapprobation; he thus
enraged the reviewer in the <u>Philosophical Magazine</u>.[25]

Davy's work on sulphur and phosphorus had
pleased Low, who wrote that Davy 'even expressed the
opinion, that all the simple bodies might be compound,
and resolvable into hydrogen and some unknown base.
He never, however, pursued his own hypothesis to its
consequences; and at length he seems to have

abandoned it altogether'. Low thought this because
in his papers on chlorine Davy had applied Lavoisier's
maxim with such rigor; he put this down to Davy's
'conversion' to the atomic theory, when in fact all
that Davy claimed was that chlorine had the same
status as other bodies described as elements — not
that it would never be decomposed.

Low, as Crookes was to do forty years later,[26]
took the 'rare earth' elements as the reductio ad
absurdum of the received view. These substances
differ extremely little from one another; they occur
only in very small quantities, chiefly in Sweden and
all together. These facts for Low 'supply us with
abundant reasons for believing them to be the
products of pre-existing matter, and not elements
distinct from one another, and from other bodies, in
their corpuscular constitution'. The same case could
be made for the platinum metals; and in general, it
was ridiculous to postulate nearly sixty elementary
bodies when two would do — or maybe even one, if we
allow it to exist in positive and negative polarity
or electric charge. This last speculation reminds
one of the view of the celebrated Lorenz Oken, in
whose Elements of Physiophilosophy — written in a fit
of inspiration — matter is derived from an inert
substratum and opposed polarities.

Low went on to criticise atomic theories in
general, for atoms cannot be verae causae; we know
nothing of their properties except by inference from
our own assumptions. Boscovich's atoms, mere point
centres of force, were manifestly the creatures not
of chemists but of metaphysicians. The atomic theory
was at best an heuristic device, valuable only when
freed from the encumbrance of useless hypothesis in
which Dalton had clothed it. If then atomism was

objectionable if taken seriously, and the prospect
of nearly sixty elements unpleasing, what was to be
done about chemistry? We note that Low gives little
help to those who might wish to apply his maxim
about the elements, whereas Lavoisier's can be put
into operational terms without difficulty. Low wrote
that hydrogen and carbon had the lowest atomic weights
and might therefore be the two real elements — again
we are reminded of Oken — or again, one or both might
be compound.[27] They were known to combine in almost
every conceivable ratio, which was a likely property
for 'real' elements; but hydrogen did not differ
from metals and radicals as much as it might be
expected to do if it were a true element. So Low
concluded that even hydrogen was probably a complex
body, and that we were unacquainted with the true
element M, which in its two states M^+ and M^- and in
different arrangements composed all matter.

Low foresaw no difficulties in accounting for
definite proportions on this view,[28] and he noted
with pleasure that most French chemists had abandoned
the term 'atom' in favour of 'molecule', although he
would have been happier still if they could have got
even further from atomism and followed Wollaston in
using only 'equivalents'. The next step was to go
through the list of elements, showing that none of
them had any right to be there. Carbon monoxide forms
an oxide and chloride just as elements do; it has
just as much right to be considered an element as
they have. Nitrogen is probably an oxide of carbon,
indeed almost certainly an isomer of carbon dioxide;
here we seem to be back amid the confusions of the
early days of pneumatic chemistry. The compound
nature of ammonium was evidence for that of the
alkali metals; potash was really $H_{17}C_5$, soda $H_{13}C_3$,

207

and ammonia H_5C_2. The alkali metals and ammonia represented a natural group as well-defined as any in biology.

For Low, similarity in properties must be a reflection of similarity in structure and in content, 'upon the principle that like effects must be due to like causes'. Using this dangerous principle as his guide, he assigned formulae similar to that of cyanogen to chlorine, bromine, and iodine. The platinum metals showed for Low no differences which could not have been brought about by heat treatment; and in general he treated all similar substances as having similar formulae, as being composed of carbon and hydrogen and ultimately of M^+ and M^-. In the realm of inorganic chemistry it was still only an inference that hydrogen and carbon were the only true elements; in organic chemistry it was already in many cases a proven fact. We should not put up with the atomic theory and the received doctrine of the elements knowing them to be fictions but waiting for something better; for that would be to remain satisfied with error. The atomic theory explained nothing that could not be otherwise explained — or accepted as a datum — and generated paradoxes about the combination of atoms that were remote from chemical facts.

In the third edition, of 1856,[29] Low claimed the support of Dumas for his views, and remarked that Faraday — to whom he had sent a copy of the second edition — had publicly committed himself to experiments on the decomposition of the elements. Low hoped that such distinguished allies would silence those 'minor chemists' who had used 'vulgar and insulting language' of his speculations, which clearly did 'fall within the fair range of scientific enquiry'. He was still

as opposed as ever to atomic theory, however useful
it might seem; finding Daltonian atomism a much-less-
satisfactory version of it than Ancient atomism:
'the notion that the parts of matter to which we
assign given weights, are not atoms but moleculae
formed of an unknown number of atoms'.

Low's book was extravagently hypothetical in
approach and polemical in tone; but it did contain
ideas which were to reappear in more distinguished
writings in the 1850s and subsequent decades. The
idea that the elements might be isomers was thus used
as a basis both for experiments and speculations during the 1840s; the experiments either — as with
Brown's — did not work, or — like Rigg's — found other
explanations; but isomerism provided chemical analogies
which radical simplifiers in chemistry needed, and in
some ways these eccentrics were nearer to the road
which led to modern theories of matter than were their
more empirically-minded opponents.

NOTES

1. M.Donovan, Chemistry,4th edn(1839)p.394; J.F.W.
Herschel, A Preliminary Discourse on the Study of
Natural Philosophy(1830)pp.307,304-5. Not only
adherents to nebular hypotheses (such as Herschel's
father) but also Galileo had tried to do what he said
astronomers did not; Two New Sciences,tr.H.Crew and
A.de Salvio(New York,n.d.)p.261; see A.Koyré,
Newtonian Studies(1965)201-20. Similarly there were
speculative geologists in Herschel's own day, like
Lyell.

2. J.H.Brooke, 'Wohler's Urea and its vital force',
Ambix,XV(1968)84-114; R.Siegfried, 'Composition, a
neglected aspect of the chemical revolution', Annals
of Science,XXIV(1968)275-93; and particularly
important,W.V.Farrar, 'Nineteenth-century Speculations
on the complexity of the Chemical Elements', British
Journal for the History of Science,II(1965)297-323;
IV(1968)65-7. C.A.Russell, The History of Valency
(Leicester,1971) descrives the later developments in

this direction. See T.H.Levere,Affinity and Matter
(Oxford,1971) chaps.5 and 6, on Berzelius and on
organic chemistry. There is an essay on Berzelius in
H.Hartley,Studies in the History of Chemistry (Oxford,
1971). F.M.Heimann, 'Faraday's Theories of Electricity
and Matter', British Journal for the History of
Science,V(1971)235-57; L.P.Williams, Michael Faraday
(1965). See D.M.Knight, 'The Vital Flame',Ambix,XXIII
(1976)5-15.

3. H.M.Leicester and H.S.Klickstein,A Sourcebook in
Chemistry,1400-1900(New York,1952)pp.264-5.

4. Leicester and Klickstein,Sourcebook,pp.306-8;
The Quarterly Journal of Science,XIV(1823)198,415;
H.Hartley,Polymorphism(Oxford,1902). Wollaston's paper
Phil.Trans.,CIII(1813)51-63, showing how crystals
could be built up from spheroidal particles was
another extension of the atomic (or really corpuscular)
theory.

5. Report of the British Association,(1837)163-215.
See the account of his life in the Dictionary of
National Biography. There is a brief account of
Johnson's paper in Athenaeum(1837)747.

6. J.J.Berzelius,Scientific Memoirs,IV(1846)240-52.

7. Report of the British Association,(1837)210. Cf
Faraday in H.Bence Jones,Life and Letters of Faraday,
2nd edn(1870)pp.225-6, and W.Ostwald,'Faraday Lecture'
J.Chemical Society,T.85(1904)521.

8. R.Kane,Elements of Chemistry,2nd edn(Dublin,1849)
pp.319,294-5.

9. J.von Liebig and J.B.Dumas,Comptes Rendus,V(1837)
567-72; H.M.Leicester and H.S.Klickstein,Sourcebook
(1952)p.325. Dumas' views will be discussed in chapter
VIII below. W.H.Brooke, 'Chlorine substitution and the
future of organic chemistry',Studies in the History
and Philosophy of Science,IV(1973-4)41-44. R.Kane,
Elements of Chemistry,2nd edn(Dublin,1849)p.324; on
this radical, see R.W.Bunsen,Scientific Memoirs,III
(1843)281; he pointed out its similarity to inorganic
atoms, and added (p.282): 'This harmony may not,
perhaps, be without some influence in a science where
every idea is supported by analogical reasoning'.
See C.A.Russell,note 2; and N.W.Fisher, 'Organic
Classification before Kekulé',Ambix,XX(1973)106-31,
209-34; 'Kekulé',Ambix XXI(1974)29-52.

10. J.von Liebig,Familiar Letters on Chemistry(1843)
pp.46-55; quotations from pp.53 and 54. On Liebig's
chemistry, see the introduction by F.L.Holmes to the

reprint of Liebig's Animal Chemistry(1964).

11. S.Brown,Trans. Roy. Soc. of Edinburgh,XV(1844)
165-76,229-49. There is an obituary of Brown in the
North British Review,XXVI(1856-7)376ff. An attack on
his procedures had appeared in Phil.Mag.,XIX(1841)
295; Brown's reply is at p.388. Brown was associated
with phrenology, and was the founder of 'itinerating
libraries'.

12. S.Brown,Lectures on the Atomic Theory, and Essays
Scientific and Literary(Edinburgh,1858)I,pp.6,10,15,
19-20,31,34. A 'mixed science' was a partly empirical
one, the'pure' sciences being mathematics and logic.
On Exley and Mossotti,and their force-centre atoms,
see D.M.Knight,Atoms and Elements(1967)chap.4.

13. S.Brown,Lectures,I,pp.17,66; Trans. Roy. Soc. of
Edinburgh,XV(1844)169,174. See G.Wilson, Edinburgh
New Philosophical Journal,XXXVII(1844)8.

14. S.Brown,Lectures,I,pp.71,72,264; cf Davy,Works,
II,p.323; T.Young,Quarterly Review,VIII(1812)84.

15. S.Brown,Trans. Roy. Soc. of Edinburgh,XV(1844)
229,236; Lectures,I,pp.78-9; see below,chapter IX,
and J.W.van Spronsen,The Periodic System of Chemical
Elements:a history of the first hundred years(1969).

16. G.J.Knox,Proceedings of the Royal Irish Academy,
II(1840-4)171; see H.Davy,Phil.Trans.,C(1810)16-74
for the Bakerian Lecture on which Knox drew.

17. See the obituary of Brown,The North British
Review,XXVI(1856-7)391; G.Wilson,Edinburgh New
Philosophical Journal,XXXVII(1844)17.

18. G.Wilson,Religio Chemici(1862); The Life of the
Honourable Henry Cavendish(1851); Electricity and the
Electric Telegraph, together with the chemistry of
the stars,new edn(1859)pp.34-6; contribution to
Edinburgh Essays(Edinburgh,1857)pp.301-50. M.Travers,
A Life of Sir William Ramsay(1956)p.100; D.S.L.
Cardwell(ed.) John Dalton and the Progress of Science
(Manchester,1968)pp.232,238-9. For Wilson's attitude
to atomism, see his Chemistry(Edinburgh,1850)pp.39-43;
Religio Chemici(1862)p.344.

19. G.Wilson,Edinburgh New Philosophical Journal,
XXXVII(1844)1-21; Transactions of the Royal Society
of Edinburgh,XV(1844)547; on paracyanogen, see J.R.
Partington,A Textbook of Inorganic Chemistry,6th edn
(1950)p.639; he suggests a formula $(CN)_n$. J.von
Liebig, Familiar Letters on Chemistry(1843)p.55; The
North British Review,XXVI(1856-7)377.

20. R.Rigg,Experimental Researches(1844); Rigg had
published papers in journals including Phil.Trans.,
and Proc. Roy. Soc. See A.Thackray,Atoms and Powers
(Cambridge,Mass.,1970)p.25, for the view that plants
transmute; S.T.Coleridge,Aids to Reflection,ed.T.
Fenby(Liverpool,1874)pp.357-8 (Conclusions,sections 9
and 10); and D.M.Knight,'Chemistry, physiology, and
materialism in the Romantic Period', Durham University
Journal,LXIV(1972)139-45.

21. Liebig's views are to be found in his Organic
Chemistry in its applications to Agriculture and
Physiology(1840); and see the paper by Charles
Daubeny,Phil.Trans.CXXVI(1836)149. R.Rigg,
Experimental Researches,pp.ix,60,186ff,119,264.

22. M.J.Schleiden,Principles of Scientific Botany,tr.
E.Lankester(1849)pp.1,2,4. Davy had made the same
point in his Elements of Agricultural Chemistry,2nd.
edn(1814)310ff; see D.M.Knight,'Agriculture and
Chemistry in Britain around 1800',Annals of Science,
XXXIII(1976)187-96.

23. D.Low,An Inquiry(1844); 2nd edn(1848); 3rd edn
(1856).

24. Phil.Mag.,XXIV(3rd series,1844)296; Low wrote that
this was merely a violent attack on him and not a
review, Phil.Mag.,XXV(3rd series,1844)116; on p.118
Richard Phillips admitted that he had written it and
promised to reply to Low's strictures, but never did
so. On Berzelius, see D.Low,An Inquiry,2nd edn(1848)
p.iv.

25. D.Low,An Inquiry,2nd edn(1848)pp.1-3,10,100,331,
342-3. B.le B.de Fontenelle,Entretiens sur la plura-
lite des Mondes,ed.R.Shackleton(Oxford,1955)p.105;
see D.M.Knight,'Uniformity and Diversity of Nature in
17th century treatises on Plurality of Worlds', Organon
(Polish Academy of Sciences)IV(1967)61-8, and
'Celestial Worlds Discover'd',Durham University Journal
LVIII(1965)23-9. G.Wilson,Edinburgh New Philosophical
Journal,XXXVII(1844)3. J.H.Brooke,'Natural Theology
and the Plurality of Worlds',Annals of Science,XXXIV
(1977)221-86.

26. W.Crookes,Chemical News,LX(1889)27-30,39-41,51-3,
63-6; this was his Presidential Address to the
Chemical Society of London. D.Low,An Inquiry,2nd edn
(1848)p.5,274. L.Oken,Elements of Physiophilosophy,tr.
A.Tulk(1847)pp.61-3; see D.M.Knight,'Steps towards a
dynamical chemistry',Ambix,XIV(1967)179-97. On
dynamical chemistry, see also M.A.Sutton,'J.F.Daniell
and the Boscovichean Atom', Studies in History and
Philosophy of Science,I(1971)277-92, and D.F.Larder,

'A dialectical consideration of Butlerov's theory of chemical structure',Ambix,XVIII(1971)26-48, and his translation of A.M.Butlerov, 'On the chemical structure of substances',Journal of Chemical Education, XLVIII(1971)287-91.

27. D.Low,An Inquiry,2nd edn(1848)pp.9-11,12-14. The work of Davy and Berzelius made electrical theories of matter very popular with speculators; see S.T. Coleridge,Hints towards the formation of a more comprehensive Theory of Life,ed.S.B.Watson(1848); and the very curious work of K.von Reichenbach, Researches on Magnetism Electricity, Heat, Light, Crystallization, and Chemical Attraction in their relations to the vital force,pts.1 and II,tr.W. Gregory(1850). The translator of this book on animal magnetism, who was a zealous proponent of its doctrines, was also the translator and populariser of Liebig's writings. Reichenbach believed that a new imponderable, odyle, lay behind all the phenomena listed in his title.

28. D.Low,An Inquiry,2nd edn(1848)pp.15,16-27. On p.28 he refers to the hypothesis of Ampere and Avogadro as 'strange' and 'opposed to various well-known facts'; see also pp.83,116-21,274,329,343.

29. D.Low,An Inquiry,3rd edn(1856)pp.vi-viii,380; by now there were more elements for Low to account for. For Dumas and Faraday, see M.Faraday,A Course of Six Lectures on the Non-metallic Elements, ed.J.Scoffern (1853)p.36; see also pp.160-66,40-43; Athenaeum (12 July,1851)p.750, for a report of the British Association meeting — see below, chapter 8. Charles Daubeny, in the second edition of his Introduction to the Atomic Theory(Oxford,1950) discussed the work of Brown,Rigg,and Low: the latter's work, he wrote, (pp.426ff,441) carried no conviction; such playing about with numbers could always be done, and 'specu-lations of this kind, even if possessed of more intrinsic probability than they can lay claim to, must be carefully kept distinct from the facts of a science like chemistry, and from the deductions that flow legitimately from them'. On verae causae in Victorian science, see the essay-review by M.J.S.Hodge, British Journal for the History of Science,X(1977)237-46.

8

The Calculus of Chemical Operations

We began the last chapter with a quotation from
Michael Donovan on Prout's hypothesis; and our
discussion of the 1850s can well begin with noticing
his paper of 1852 on the supposed agent responsible
for the phenomena of electricity and magnetism, for
he reminds us that simplicity is a matter of taste:
'if we diminish the number of agents, we must
increase the number of their properties, in order to
explain, with any degree of probability, the diversity
of natural phenomena; hence nothing is thereby
simplified, and no advantage gained'.[1] Almost everything in nature, on Donovan's view, from minerals
through the biological realm to air and light, is
compound and complicated, and different phenomena are
probably produced by different agencies. But this
argument seems to have fallen on deaf ears, coming as
it did just as the principle of conservation of energy
was being used to bring together sciences that had
seemed distinct, and create our 'physics'.

In such a climate of opinion, belief in the unity
of matter began to seem once more respectable.
Support for it was forthcoming from such eminent
chemists as Dumas and Hermann Kopp, as well as from
others well-known and highly-placed in their day
though now little remembered. Dumas had played a

leading part in theoretical chemistry and had the
good fortune to survive to a ripe old age, outliving
his opponents. In 1860 at the Karlsruhe Conference[2]
he received an ovation as the greatest living chemist,
and in 1869 he delivered the Faraday Lecture to the
Chemical Society in London; this is a source of his
most mature views on the nature of the elements. The
redoubtable philosopher of science J.B.Stallo, a
German American, referred in 1881 to

> the ceaseless recurrence among
> French physicists and chemists
> of the attempt to exhibit the
> chemical elements as compounds
> or allotropic forms of some single
> element [which] afford striking
> proof of the logical constraint,
> resulting from the atomomechanical
> theory, to reduce the primordial
> atoms to uniformity and equality.

This sally was particularly directed at Marignac,
whom we shall meet shortly; it was unfair to single
out the French for their devotion to the unity of
matter, but it may be that the corpuscular tradition
of Laplace and Berthollet lived on into the epoch of
Dumas and of Berthelot, and that French chemists were
better Newtonians than their British or German contemporaries.

Dumas in his lecture of 1869 asked the questions
which are now familiar to us; are the elements complex?
and were Lavoisier and Dalton between them the Kepler
of chemistry, which still awaits its Newton? Elements
as defined by Lavoisier, he said, had all the same
status;[3] they were 'substances of the same order, and
if we may say that they cannot be resolved, it is
only because they have not been resolved'. He
referred to the three standard arguments for the
complexity of the elements; their number was

excessive, and increasing; their natural classification by families pointed to common constituents; and
their analogies with the radicals of organic
chemistry, which were known to be compound, indicated
that they might well be themselves compound radicals.
Dumas had in 1851 addressed the British Association
at Ipswich and urged this last argument; upon which
Faraday expressed the hope that he 'was setting
chemists in the right path' and other eminent
chemists and physicists concurred. Dumas had then
even spoken of the possibility of transmutations.

In 1869 Dumas recognised that there were arguments on the other side. No element had ever been
observed to break up; their atoms occupied the same
space in the gaseous state as in the solid, in which
related elements could replace one another in isomorphous bodies; and their atomic heats were all
approximately identical. They were thus a fairly
well-defined class; but this did not prove that they
were not compound. Since the definition of an
element was that it could not be decomposed, it was
not surprising that none had been; but Dumas avoided
this circle by his urging that the elements were all
members of a class, having similar status, and that
therefore the decomposition of one would cast doubt
upon the simplicity of all. Dumas did not in this
paper refer to the controversy that had been in
progress among his disciples on the question of the
truth of Prout's hypothesis: in 1860 Stas published a
resumé of his numerous determinations of atomic
weights, concluding as Berzelius had that Prout's
view could not stand;[4] and in the same year Marignac
published a paper drawing the opposite conclusion.
These papers had some bearing on the problem raised
by Dumas, of the strength of the analogies between

217

radicals and elements. At the British Association in
1851 we are told that:

> Prof. Dumas had not previously
> prepared diagrams or tables, but
> covered a large blackboard with
> lines, figures, and formulae, to
> follow his train of reasoning, —
> and symbols, volumes, and names
> were rapidly produced and as
> rapidly effaced to illustrate the
> Professor's views of the laws of
> the substitution of one body for
> another in a compound.

But he went on in succeeding years to embody his
ideas in formal papers; notably in one of 1859, which
reminds us of Thomson's First Principles, though Dumas
was able to avoid some of the worst traps into which
Thomson had blundered. There are, Dumas declared, two
distinct views of the elements; either they are
distinct bodies, as Berzelius had held, or they are
built up by the condensation of prime matter in
different arrangements. In Dumas opinion, Berzelius'
analyses had by no means eliminated the latter
possibility, and Prout's hypothesis was still unfalsi-
fied; for Berzelius had been unable to explain the
numerous coincidences on the list. The numbers were
not a random selection; many were approximately
integer multiples of that of hydrogen, and several
were simple multiples of one another.

Believing that such coincidences could not be
fortuitous, Dumas asked four questions. The first
was whether the equivalents of all elements were
integer multiples of that of hydrogen — 'equivalents'
because Dumas was sceptical about Daltonian atoms. To
this question Thomson had given an affirmative answer,
and Berzelius and Turner a negative; and Dumas was
careful. Chlorine was the crucial case; his determinations of its equivalent gave a mean value of 35.5 .

Other elements also proved to have fractional
equivalent weights, notably barium, manganese, and
nickel; but these fractions seemed to be exactly
halves or quarters. Prout, we remember, had suggested that hydrogen itself might be built up from simpler
prime matter; and if to this prime matter, not
encountered in isolation, were assigned an equivalent
weight one quarter that of hydrogen, then a version of
Prout's hypothesis could be saved, and indeed regarded as verified.

The second question concerned the ratios between
atomic weights. Dumas performed series of experiments
determining the equivalents of chemically related and
unrelated elements, and a number of simple ratios did
appear. Thus oxygen and sulphur stood in the ratio
2:1, and nickel and cobalt shared the same value;
these pairs of elements belong to the same natural
groups, but nitrogen and iron, and iron and cadmium,
were also in the ratio 1:2 although there were no
close chemical analogies between them. On the other
hand, the very similar metals molybdenum and tungsten
persisted in the ratio 48:92 in spite of all Dumas'
efforts to make it 1:2.

The third question concerned the triads of
related elements, like chlorine, bromine, and iodine.
Here Dumas asked whether the equivalent weight of the
middle member was exactly the mean of the others; but
extensive work indicated that in some cases it was
and in others it was not. This third question led
into the fourth, whether the elements were close
analogues of the organic radicals or not. In the
homologous series of similar radicals in organic
chemistry — for example the series which begins
methyl, ethyl, propyl, — the equivalent weight of the
nth member can be represented by a formula of the

type ($a+nd$), where $\underline{a}$ is the equivalent weight of the
first member and $\underline{d}$ a constant. The halogen group —
chlorine and its congeners — could not be represented
by a formula of quite this simplicity; but the group
carbon, boron, silicon, and zirconium fitted tolerably
well the series $\underline{a}$, $\underline{a+d}$, $\underline{a+3d}$, $\underline{a+12d}$. We should
perhaps note that modern chemists do not put all
these elements in the same group, as Dumas did. He
found some correlation between the weights of
corresponding elements in the different families.
Thus comparing the nitrogen family with the halogens,
he found that each halogen weighed five units more
than the corresponding member of the other family.
Similarly, the oxygen family ran parallel with the
alkaline earths starting with magnesium, the difference being four units; a similar parallelism, with a
difference of three units, was observed between the
homologous series beginning with methylene, and the
amine series beginning with ammonia.

Such results, in Dumas' opinion, justified doubts
as to the ultimate irreducibility of the elements,
and indeed indicated that we could expect their decomposition. But naturally they did not cast any
light on the manner in which this decomposition might
be brought about. Lavoisier had according to Dumas
shown a happy mixture of boldness and prudence in his
remarks and conjectures on the elements, and like all
masters had not confused the limits of analysis with
the true elements of matter.

That Dumas did not in 1869 mention these engaging
quantitative speculations is a tribute to the analyses
of J.B.Stas.[6] For Stas played in relation to Dumas'
resucitation of Prout's hypothesis the part which
Berzelius and Turner had played on its appearance in
Thomson's First Principles. Stas did not simply set

out to prove that the atomic weights of the elements
were not whole numbers relative to that of hydrogen,
or a simple fraction of hydrogen; having found out
that they were not, he concluded that the elements
were not complex. This was a step which Turner, for
example, had refused to take. Stas wrote that

>from the point of view of the
>philosophy of nature the import of
>Prout's idea is immense. Those
>components of compound substances
>which we regard as elementary on
>account of their absolute immut-
>ability for us, would themselves
>be only compounds. Such elements,
>whose discovery is the glory of
>Lavoisier and has immortalised his
>name, can thus be considered as
>being derived from the condensation
>of a unique substance: we are thus
>naturally conducted to the unity of
>matter, although we in fact observe
>its plurality, its multiplicity.

Stas and Dumas had determined the atomic weight of
carbon, and found it to be an integer multiple of that
of hydrogen; and when he began his researches Stas had
been confident in the truth of Prout's hypothesis in
one of its forms.

Stas found that different experimental procedures
led to discordant results, but by following (like
Berzelius) accepted and well-tried methods he achieved
better control; and he based his final values not only
upon analysis but on syntheses and double-decompositions. By 1860 he could report that he had 'reached
the complete conviction, the entire certainty, as far
as certainty can be attained on such a subject, that
Prout's law, with all the modifications due to M.Dumas,
is nothing but an illusion, a pure hypothesis express-
ly contradicted by experiment'. Indeed, if oxygen
were assigned the equivalent weight 8, then the
equivalent of hydrogen was not exactly unity; so the

very foundation of the Proutian scheme was infirm. Stas' experiments, with all their precautions for the avoidance of error, duly became classics; and he concluded that 'as long as we hold to experiment for determining the laws which regulate matter' we must consider Prout's hypothesis an illusion, and 'regard the undecomposeable bodies of our globe as distinct entities having no simple relation by weight to one another. The incontestable analogy of properties observed amongst certain elements must be sought in other causes than those originating in the ratio of weight of their reacting masses'.[7]

It will cause no surprise to find that this paper did not completely stem the tide of Proutian speculation; an hypothesis so attractive could survive several falsifications. Many chemists, among them Dumas, had reached a belief in the complexity of the elements along a different path; if the Proutian approach proved a dead end, there remained the arguments marshalled by Dumas in the Faraday Lecture. But it was even possible to claim that Stas' results had not disposed of Prout's hypothesis; that all that was necessary was to round off a few of Stas' figures, and to take a smaller fraction of hydrogen as the prime matter. This was the course followed by J.C.G.Marignac

Marignac compared the results of Stas' analyses with those predicted in accordance with Dumas' modification of Prout's Hypothesis. He found that while they were quite near, there were divergencies too great to be put down to experimental error. Marignac admitted that Stas' results were considerably more accurate than those he had himself obtained earlier, and yet were no nearer the values predicted by the Proutian theory:

so,if some chemist at a future date
takes up the same series of
experiments with still greater
guarantees of accuracy, the difference
which may be found between his
results and those of M.Stas will
probably be of the same order as
that which exists between the latter
and my own, and ... no greater
accordance with Prout's law will be
found.[8]

Having made these admissions, he refused nevertheless
to accept that Stas had refuted Prout. Despite the
remarks just quoted, he urged that chemists should
make atomic weight determinations by a whole series of
methods, beginning with a wide range of starting
materials; while such analyses would be very laborious,
they might lead to values more in accordance with
Prout's view.

Even if one rejected this, Marignac urged that
one must admit that many atomic weights are very close
to whole numbers. This could not be accidental; and
Prout's law should be viewed perhaps as one views the
gas laws, as accurate for practical purposes, and
probably completely if 'disturbing influences' could
be computed and eliminated. The law thus expresses
'perhaps ... the normal ratios which should exist
between these weights, apart from some perturbing
causes the elucidation of which may in future exercise
the sagacity or the imagination of chemists'. One
could say that this prediction was verified in the
'packing fraction' of Aston, and indeed Marignac did
suggest that the manner in which the prime atoms were
compressed into elements might affect the way gravity
acts upon them. But he had none of the evidence of
the existence of isotopes available to Aston sixty
years later, and his suggestions were vague. We
should take him to be making a logical point about

223

the exactness of laws rather than to be proposing a physical explanation of why atomic weights display such regularities as they do display.

He made a further logical point in reply to Stas' remark that since the equivalents were not whole multiples of that of hydrogen, or a vulgar fraction of that of hydrogen, the elements must be simple. Marignac urged that Prout's fundamental idea of the unity of matter was independent of the magnitude of the divisor which expressed the weight of the atom of 'protyle'. The weight could be one half that of an atom of hydrogen, or it could be a much smaller fraction — say one thousandth. If the latter figure were adopted instead of the former, all that would change would be the complexity of the known elements. This move would enable Proutians to be always one step ahead of the analysts; but in the absence of any direct evidence for protyle it does not seem a very happy line to pursue. But there is again an analogy in the history of astronomy. The Copernican theory implied that the fixed stars must be an extremely long way off because their relative positions did not change as the Earth moved around its orbit; these observations were in a sense an experimental test of the Copernican theory, which it failed — and yet survived. As telescopes were improved, still no stellar parallax was observed and it became necessary each time to suppose the stars still further away. By the time stellar parallax was detected by Bessel in the 1830s everybody was a Copernican anyway. The minute particle which J.J.Thomson described as a 'corpuscle' in his classical paper of 1897[9] and which he believed to be the prime matter — our electron — might with some exercise of generosity be held to be that, the possibility of which was predicted by

Marignac.

While these experiments were being performed in France, the questions which Dumas had raised at Ipswich in 1851 were asked in Presidential Addresses at subsequent meetings of the British Association. Thus Charles Daubeny, who was President in 1856, declared that Davy might 'at the close of his career, have despaired of adding anything worthy of his name to the domain of chemistry, and sighed for other worlds to subdue'. But the rise of organic chemistry had shown the importance of the arrangement of the component elements in a compound; and the persistence of the compound radicals had suggested 'that many of the so-called elements of inorganic matter may likewise be compounds, differing from the organic radicals above mentioned merely in their constituents being bound together by a closer affinity'. The existence of the triads of related elements confirmed this hypothesis:

> when we observe a series of bodies,
> exhibiting, as it would seem, a
> gradation of properties, and
> although as yet undecompounded,
> possessing a common numerical
> relation one to the other, who will
> deny the probability, that they are
> composed of the same constituents,
> however little approach we may have
> as yet made towards the art of
> resolving them into their elements,
> or of forming them anew?[10]

Twenty-five years before, in his book on the atomic theory, he had been more cautious.

In 1858 the eminent palaeontologist Richard Owen remarked in his Address that 'the present tendency of the higher generalizations of chemistry seems to be towards a reduction of the number of bodies which are called 'elementary'; it begins to be suspected that

225

certain groups of so-called chemical elements are but modified forms of one another'.[11] In the following year the chemist Lyon Playfair informed his Section of the British Association that Sir John Herschel had regarded the decision for or against Prout's hypothesis as crucial, and that Dumas' results now seemed to provide a basis for such a decision: 'the general view of Prout, that the equivalents of the elements, compared with certain unities, are represented by whole numbers, seems to be established by recent experiments, though it would be premature to declare that there are no exceptions to the law'. None of these ideas were particularly new, but suggestions thus mooted from the chair at the British Association were bound to gain wide currency; and also give an indication that the idea was not regarded as wild or fanciful.

In 1865 a new approach to the problem was indicated in a very long paper on specific heats by Hermann Kopp, which appeared in the <u>Philosophical Transactions</u>; English readers had had two years earlier a brief paper on the same subject in <u>Chemical News</u>. The original point of these researches was to find a way of determining atomic weights from equivalents; for chemists in the 1850s had often used these terms as synonyms, despairing of getting agreement on genuine atomic weights. The Karlsruhe Conference of 1860 met to discuss this situation, and the disagreements over chemical formulae to which it had given rise; it accepted after a vote that equivalents were more empirical than atoms or molecules. Analyses gave equivalent weights; to convert these to atomic weights it was necessary to have some indication of how many atoms there were in the compound analysed. The law of isomorphism could be used as a guide, but compounds were known which were isomorphous although they contained different numbers of atoms; for example, potassium and ammonium salts. Another method was to use Avogadro's hypothesis that equal volumes of all gases contained, under the same conditions, equal numbers of molecules; but in 1860 this had fallen into disrepute although independent evidence for its truth was forthcoming from the kinetic theory of gases. At Karlsruhe Cannizzaro tried to resurrect it, but scant attention was paid to him — fortunately he had prudently had offprints of his paper prepared, and circulated; it was not until they returned home and read his arguments that many chemists came to accept them.[12]

There was a third approach to this problem, based upon the law published in 1819 by Dulong and Petit, that the product of the atomic weight and the specific heat of elements is a constant, which they called the atomic heat.[13] The law, which seems to have been purely empirical and was independent of the form of the substance, they expressed in the form: 'atoms of all simple substances have the same capacity for heat'. On the usual atomic weight scale (H=1), the value of the constant was about 6.4; the law thus led to a rough atomic weight, and an accurate one could be computed from the equivalent.[13]

To Kopp the idea occurred that it might be possible, given that atomic weights could be determined by other means, to use Dulong and Petit's law as a test of elementary status; for if the atomic heat were not about 6.4, then the substance would not be an element.[14] His paper of 1865 began with an historical review; and contains full descriptions and results of numerous calorimetric observations. He found that the law failed badly to fit the non-metals; thus for

sulphur and phosphorus, the 'constant' turned out to
be 5.4; for fluorine, 5; for oxygen, 4; for silicon,
3.8; for boron, 2.7; for hydrogen, 2.3; and for carbon,
1.8. The property seemed to be additive, the 'atomic'
heats of compounds being the sum of those of their
components.

In the sixth part of his paper Kopp turned to
the question of the elements, asking: 'what bodies
are to be regarded as chemical elements? Does the
mere fact of indecomposeability determine this? or
may a body be indecomposeable in point of fact and
yet from reasons of analogy be regarded not as an
element but as a compound?' The history of chemistry
— and Kopp was an historian — showed that affirmative
answers had been given at various times to both these
questions; Lavoisier had written that the alkalies
and earths must be deemed compound, from analogy, and
Davy had raised chlorine to elementary status because
no experiment had ever decomposed it. According to
Kopp, Berzelius had fallen in with Davy's view of
chlorine 'more ... because he was outvoted than
because he was convinced', and Schönbein was well-
known as still believing, on grounds of analogy, that
the halogens were oxides.

Had Dulong and Petit's law proved universally
valid, it would have served as a test of elementary
nature; at least it would have indicated that all the
elements were of the same order of complexity. But
in the event, it had turned out that it was possible
for a compound body to have the same 'atomic heat' as
an element; the value for hydrogen peroxide, for
example, is 6.3. The greater 'atomic heats' of
cyanogen and ammonium compounds might indicate that
they were more complex than corresponding derivatives
of chlorine and potassium, for greater simplicity

could be presumed to be correlated with lower atomic
heat:

> carbon and hydrogen, for example,
> if not themselves simple bodies,
> are more so than silicium or oxygen;
> and still more complex compounds
> are the elements which are now
> considered as following Dulong and
> Petit's law ... One might be
> tempted, by comparing atomic heats,
> to form an idea how the more
> complex of the present indecompose-
> able bodies might be composed of
> more simple ones.

However more accurate experimental results are
necessary first.

Kopp's speculations ended with the sentence: 'But
even if these suppositions are not free from uncertainty and imperfection, they appear worthy of attention
in a science which, for science, is still so much in
darkness, as is the nature of the indecomposeable
bodies'. As it happened, Kopp's arguments, like those
based upon the analogies between the elements and the
compound radicals, did not prove to be well-founded,
and the speculative part of his paper proved a cul-de-
sac. But speculations which did find their way into
modern chemistry arose from the application of
spectroscopy in astronomy at about the same time.

The dark lines in the solar spectrum had been
noticed by Wollaston and studied in detail by Fraunhofer. The work of Bunsen and Kirchhoff on spectrum
analysis followed;[15] and then Kirchhoff — and Stokes
independently — suggested that the dark lines might
be due to the presence in the Sun's atmosphere of
bodies which would give bright lines at the same
frequency in their arc spectra. Some spectroscopists
were at first dubious about this identification of
emission and absorption lines, particularly as the

latter had not been produced in the laboratory; for example Crookes in 1861 voiced his disquiet. But the majority were convinced, and in 1864 the chemist W.A. Miller and the astronomer William Huggins published papers on the spectra of heavenly bodies. They had used a refractor with an eight inch objective, and two dense flint glass prisms in their spectroscope — the two prisms give a high dispersion, weakening the continuous spectrum, but the lines are separated further and stand out just as strongly. No stellar spectrum appeared devoid of lines, and in some fixed stars a number of elements could be identified; nine of them in Aldeboran. They concluded that chemically stars seemed to be very like the Sun, and suggested that probably therefore they had planets encircling them. They seemed thus to have disposed of the engaging hypothesis of George Wilson that since there was no reason why there should be, for example, so much iron and so little gold in the world, elements which are rare on the Earth's surface must be common somewhere else.

They were also instrumental in reviving the nebular hypothesis of the origin of solar systems; for with better telescopes, particularly that of Lord Rosse so many nebulous objects had been resolved into star-clusters that it seemed reasonable to suppose that all would be in time.[16] But in the spectrum of a nebula Miller and Huggins observed a bright line due to nitrogen, indicating that they were looking at a hot gaseous object and not at a star. They found, however, that some nitrogen lines were not to be seen although they should have been obvious. Huggins suggested that this might indicate that in the nebula there was a form of matter more elementary than nitrogen; the spectrum of the element consisting of lines

contributed by its various components. Such speculations were followed up by Norman Lockyer (the founder
of the journal Nature, and a greater astronomer bold
in his conjectures) who adopted the working hypothesis
that the chemical elements were compound bodies, and
provided spectroscopic evidence for this view particularly from his observations of the Sun.

This had some effect on the speculations of
chemists, although the majority of them were unwilling
to attach much importance to physical evidence at this
period. In particular, their work was referred to by
Benjamin Brodie[17] who followed Daubeny as Professor of
Chemistry at Oxford from 1855 to 1872; by this time
there was an undergraduate course in the subject, and
Brodie introduced laboratory instruction. He was the
son of the famous physiologist of the same name, who
had been President of the Royal Society; and he was
among the first chemists to make the pilgrimage from
England to Liebig at Giessen for training in research.
He began as an atomist, and his first major paper was
directed against Berzelius' dualism; he sought to show
that rather than having a definite electrical state
the atoms of the elements become polarised during
reactions. He drew attention to the way in which
chemists wrote of the particles of elements as being
held together only by the force of aggregation, while
those of compounds were retained in place by the force
of affinity. Berzelius' electrical theory made it
impossible that like atoms,being similarly charged,
could form chemical links; elements must therefore, on
the received view, consist 'of single and isolated
atoms'.

Brodie tried to show, in contrast to this view,
that the elements did not seem to be a distinct class
of bodies as this would imply, and that many reactions

involving elements can be best understood if we
assume that elements behave not unlike compounds; he
therefore tried to extend the electrochemical theory
so as to allow for chemical links between similar
atoms. As reaction begins these links become polarised so that when for example iodine reacts with
baryta one atom of iodine has induced in it a positive
charge while the other is negative; they can therefore
combine with the barium and oxygen atoms in a double
decomposition which he wrote: $\overline{I^+I^-\ Ba^+O^-}$ = IBa + IO.
He concluded: 'I have limited my assertions as to the
nature of the elemental bodies, to the statement, that,
under certain conditions, there exists a chemical
difference between the particles of which they consist.
On the chemical nature of these particles, I have
offered no opinion'. He referred to the theory of
Ampère, that the elements consist of like particles
in a state of combination; but remarked that though it
seemed the 'apparent' interpretation, it was not
necessarily the most 'rational and philosophical' view,
and indeed ran us into logical and theoretical
difficulties. But besides the views of Berzelius and
of Ampère, there was

> yet a third hypothesis which we may
> form as to the constitution of these
> elements, namely, that they consist
> of yet other and further elements.
> On this view, the real fact which lay
> hid under these phenomena, might be
> the synthesis of oxygen (or other
> elements) from the ultimate and
> farther elements of which the oxygen
> consisted. In the present state of
> our knowledge it is useless to dilate
> on this idea; but on the assumption
> that the elemental bodies are in this
> sense compound, we may, I believe,
> account for all these experiments
> without ever assuming a chemical
> difference between two similar

232

particles; and it appears to me
perfectly possible, that in such
a constitution of the elemental
bodies, these experiments, together with the phenomena of
allotropy, may find their
ultimate solution.

Polarisation is naturally easier to explain in complex
molecules than in hard indivisible atoms, or in
molecules of like atoms somehow fixed together; but
Brodie's argument, we might note in passing, demands
more than one kind of prime matter.

Brodie published other papers,[18] on the organic
peroxides which he compared to halides, and on ozone;
it is for this that he is still remembered in histories of chemistry. He had also investigated the allotropy of carbon, concluding that in graphite it
behaved as a distinct element of atomic weight 33
which he called graphon. His attention was next
called to the problem of chemical notation. This,
partly because of the confusion between atomic and
equivalent weights and partly through divergencies of
theory, had become a scandal. In addresses to the
British Association eminent scientists called for a
reform; and in 1864 William Odling,(the translator of
Laurent, who had proposed the hypothetico-deductive
method for determining molecular structures) as
President of the Chemical Section urged Brodie to
publish his 'new and strictly philosophical system of
chemical notation by means of actual formulae instead
of mere symbols'.

We need not pursue here the details of Brodie's
operational calculus, which was announced to the Royal
Society in 1866 and described in a more popular manner
to the Chemical Society in the following year.[19]
What is interesting to us is its bearing upon the

question of the nature of the elements. Brodie's
system was designed so that the question whether
atoms existed, and a fortiori how they were arranged,
could not be raised. Substances were regarded as the
result of an operation performed upon a unit of space;
α symbolised the operation by which a unit of space
became a unit of hydrogen, and other Greek letters
stood for similar operations, which could not be
further investigated. Some substances seemed to
require only one operation to make them, but others
could be regarded as composed in two or more similar
or different operations; thus water is $\alpha\xi$ and oxygen
ξ^2 . The symbols obey a Boolean algebra; Brodie's
symbols thus avoided the objection made by Herschel
that for example the formula HCl looks as though it
means hydrogen multiplied by chlorine. Brodie's
symbols obeyed clear rules, and with the aid of these
rules and an assumption of simplicity one could construct the formulae of substances from empirical data.
The formula for water, representing the operations
upon space necessary to transform it into water, was
deduced from the data on the combination of oxygen
and hydrogen; and the simplest formula from the
infinite range possible was selected. To each
operation a weight corresponded; when the operation
α had been carried out, the unit of space would have
been transformed into a unit having weight 1; if ξ
were then carried out, the result would be water of
weight 9.

This curious scheme was used by Brodie in his
studies upon ozone, but it never seems to have been
used by anybody else. It was criticised in the
discussion[20] which followed the lecture of 1867 by
G.G.Stokes, who suggested that while the calculus
took into account existing knowledge it would be

234

liable to disruption with every new discovery, which might entail the recalculation of all the formulae. Stokes, with William Thomson, had always been a great advocate for the view that science was more than a matter of calculi which fitted the facts; indeed he had written that 'a well-established theory is not a mere aid to the memory, but it professes to make us acquainted with the real processes of nature in producing observed phenomena'. For the Cambridge school of physicists of which he was a founder member it was an axiom that physical reasonableness as well as mathematical elegance was to be demanded of any theory; and Brodie's system, in which all attempt at explanation or physical reasonableness was eschewed, could not but seem retrograde. Other objections were made by chemists, who pointed out that Brodie's simplicity rules were very similar to those to which he objected in the ordinary scheme, and that his formulae could be readily transformed into the ordinary atomic notation of a one-volume system; that is, one in which the molecule of hydrogen is, in our terms, assumed to consist of one atom.

In fact Brodie did not allow his own rules against speculation to tie his hands completely.[21] His formulae were on his own admission 'ideal', not empirical yet not imaginary, but having a status like that of the entities of geometry. They were derived according to simplicity rules, and were intended to give information about the relative weights of substances, and to act as condensed recipes; but not to give any information about such hopelessly-conjectural matters as the structure of compounds. His formula for chlorine came out as $\alpha\chi^2$. Nothing should follow from this; if he had taken hydrogen to be α^2 instead of α (a two-volume system, like that which we use), then the formula of chlorine would have come out as χ^2 ; and various other symbols would also have changed. This should not matter, for the symbols have been chosen to give the simplest formula, not to represent some state of affairs.

But Brodie noticed that the formulae fell into three classes. Some represented one operation; others an operation twice repeated, like ξ^2 for oxygen; and others more than one kind of operation, like $\alpha\chi^2$. No substance had assigned to it the formula χ or χ^2 ; we have seen that this was a consequence of Brodie's formalism, but he allowed himself to believe that the substance χ might have existed in the primeval Earth, and might be found in stars and nebulae. In this connexion, Brodie drew the attention of his audience to the researches of Miller and Huggins. He also remarked that of the elements identified in the Sun proportionally very few were of his third type, made in a number of different operations, and suggested that perhaps the Sun might be too hot for them to be formed. About ten years later he became excited at the report that Victor Meyer had succeeded in decomposing chlorine, and getting oxygen out of it, although this was not a product which Brodie's formula would have led one to expect.

Brodie's prediction that some elements were complex seems to have aroused more interest among chemists than did the details of his operational calculus, which few chemists can have been equipped to understand. We should notice that as in his earlier discussions of polarity among the elements, Brodie did not envisage the unity of matter. Some elements had symbols which indicated that they were complex, but even if all these could be analysed there would still be a long list of true elements. The

236

heavier elements had on the whole the most complex formulae, as might be expected; these are of course less abundant in the Sun than in the Earth's crust.

Brodie's startling approach to chemistry threw the atomists into some consternation; but as valency theory was being developed, and the cash-value of structural considerations in organic chemistry became increasingly obvious in the work of the stereochemist, it became clear that there was no need to take the calculus of chemical operations seriously.[21] Brodie asked for time to try to account for the facts of isomerism — a phenomenon readily accounted for in terms of arrangements of particles — but he died without having completed his system. At the British Association in 1867 Anderson, President of the Chemical Section, expounded the calculus. He pointed out that Ampère's idea 'that the chemical atom might with advantage be considered to be a congeries of smaller particles' had not been taken seriously because it was purely speculative. Brodie's calculus, he went on

> involves a large amount of hypothesis,
> for it assumes that a considerable
> number of those substances hitherto
> regarded as elementary are really
> compounds. I do not imagine that
> much difficulty will be experienced
> by anyone in admitting this, for I
> apprehend that there is no chemist
> who imagines those bodies we call
> elements to be the ultimate
> constituents of matter, or who doubts
> that the time, though still far
> distant, will come when they may be
> resolved into simpler substances.

But he added that there was no empirical evidence for the view that chlorine contained hydrogen, as Brodie's symbol $\alpha\chi^2$ for it implied. Henry Roscoe, in his

obituary of Brodie, also took the line that the
calculus was really a device for predicting which
elements would be decomposed, and remarked 'no
experimental evidence was offered by him, and none
of a satisfactory character was forthcoming'.

It is curious that what had begun as an attempt
to purify chemistry from hypothesis should have been
taken as itself an hypothesis about the nature of the
elements; Brodie should have been more careful if he
wanted his calculus to be studied seriously, but by
thus contributing to an old but active controversy he
made sure that his views attracted attention. Those
who wished to get rid of hypotheses in chemistry in
the years after Brodie tended to turn to the active
and prestigious science of thermodynamics.[23] The
Phase Rule provided an example of a lead into chemistry
in which no hypotheses about the various components
were involved; and isomerism might have proved amenable
to treatment in terms of the different quantities of
chemical energy in the various isomers. An example of
this way of looking at the phenomena of chemistry is
the Faraday Lecture of Wilhelm Ostwald, delivered in
1904 soon after the discovery of radioactivity had
shown the instability of some atoms. He argued that
matter is simply bounded force, and that there is no
reason to suppose it atomic or granular. The
chemical elements are simply regions in which the
energy is at a minimum; usually to transform one to
another requires the crossing of a high potential-
energy barrier; but in the neighbourhood of uranium
the barriers are low, and spontaneous disintegration
possible.

This has taken us far from the main theme of this
chapter, which was the refusal of many chemists — even
highly-important ones — to accept that there was a

great gulf between the elements and compounds. The
analogies between elements and compound radicals led
many if not most of those who studied them to believe
that the elements were in fact stable radicals.
Quantitative forms of this hypothesis began with
Prout, whose theory seemed to have been overthrown by
Berzelius and Turner. Nevertheless, it was revived
by Dumas, and again overthrown, this time by Stas.
Marignac continued to believe that there must be something in it, and the researches of Gladstone and de
Morgan showed that the distribution of atomic weights
was far from random.[24] In 1880 Balfour Stewart wrote
that impurities might have vitiated Stas' work, which
had not disproved Prout's hypothesis; and in 1882 Lord
Rayleigh set out on the analyses of nitrogen which
began as a test of Prout's hypothesis and ended in
the discovery of argon. In the latter years of the
nineteenth century, belief in the complexity of the
elements was sustained by the chemical analogies with
which we are already familiar; by arguments from
spectroscopy similar to those we have met in Miller
and Huggins and in Lockyer; and by theories of inorganic evolution, which in a Darwinian epoch carried
all before them. These beliefs fitted in most
opportunely with the achievement in the 1860s of a
satisfactory system of classification for the elements.
Chemistry may still have been awaiting its Newton; but
in Mendeleev it found its Linnaeus, and it is to
chemical taxonomy that we must turn in our final
chapter.

NOTES

1. M.Donovan,Phil.Mag.,III(4th series,1852)117-27;
H.I.Sharlin,The Convergent Century(New York,1966);
T.S.Kuhn, in M.Clagett(ed.) Critical Problems in the

History of Science(Madison,Wis.,1959)pp.321-56.

2. On the Karlsruhe Conference, see Chemical News, II(1860)226; H.Hartley, 'Stanislao Cannizzaro and the first international chemical conference at Karlsruhe in 1860',Notes and Records of the Royal Society,XXI (1966)56-63; this is not the first international conference — see M.P.Crosland, 'The congress on definitive metric standards,1798-1799:the first international scientific conference?',Isis,LX(1969) 226-31. J.B.Dumas,Chemical News,XX(1869)1-20; J.B. Stallo,The Concepts and Theories of Modern Physics, ed.P.W.Bridgman(Cambridge,Mass.,1960)p.27; R.Meyerson, Identity and Reality(1930)p.239.

3. J.B.Dumas,Chemical News,XX(1869)3; The Athenaeum (12 July,1851)p.750; S.C.Kapoor, 'Dumas on organic classification',Ambix,XVI(1969)1-65.

4. Alembic Club Reprints,XX(1932)41-58, reprints the papers, in translation, of Stas and Marignac.

5. J.B.Dumas,Annales de Chimie et de Physique,LV (1859)129-210; especially pp.133,141,160,201. See also Alembic Club Reprints,XX(1932)14; C.Daubeny, An introduction to the Atomic Theory(Oxford,1831)p.129, for the views of Prout. For the view that the chemist should be both bold and prudent, see C.T.Schneer, Mind and Matter(New York,1969)pp.274-6.

6. J.B.Stas,Alembic Club Reprints,XX(1932)42; J.S. Stas and J.B.Dumas,Comptes Rendus,XI(1840)991-1008; Annales de Chimie et de Physique,I(1841)5-38.

7. J.S.Stas,Alembic Club Reprints,XX(1932)45,47.

8. C.Marignac,Alembic Club Reprints,XX(1932)53,57.

9. J.J.Thomson,Phil.Mag.,5th series,XLIV(1897)311. On stellar parallax, see H.Grant, History of Physical Astronomy(1852)pp.544-53.

10. C.Daubeny,Report of the British Association(1856) xlviii; cf his Introduction to the Atomic Theory (Oxford,1831)p.39.

11. R.Owen,Report of the British Association(1858)lx; L.Playfair,Report of the British Association(1859)65; J.F.W.Herschel,Preliminary Discourse(1830)p.307. There is a 'Proutian' paper by Playfair in Chemical News,II (1860)1-3; and a curious paper by J.G.Macvicar,Report of the British Association(1859)72.

12. H.Kopp,Phil.Trans.,CLV(1865)71-202; Chemical News, VIII(1863)90-1. On the Karlsruhe Conference, see note 2 above, and S.Cannizzaro,Alembic Club Reprints,XVIII (1961).

13. H.M.Leicester and H.S.Klickstein,A Sourcebook in Chemistry 1400-1900(New York,1952)pp.272-5. See M.P. Crosland,The Society of Arcueil(1967)pp.406-7; R.Fox, 'The Background to the Discovery of Dulong and Petit's Law',British Journal for the History of Science,IV (1968)1-22.

14. H.Kopp,Phil.Trans.,CLV(1865)71,199,201,202.

15. See W.McGucken,Nineteenth Century Spectroscopy (Baltimore,Md.,1969); M.A.Sutton has written (1974) an Oxford D.Phil. thesis on aspects of nineteenth century spectroscopy; see also his paper 'Sir John Herschel and the development of Spectroscopy',British Journal for the History of Science,VII(1974)42-60. Kirchoff's paper is in Berlin Monatsberichte(1859)662-5,783-7; tr. in part in W.F.Magie,A Sourcebook in Physics (New York,1935)pp.354-360; see also Nature,XIII(1876)118-9, on Stokes' precedence; and W.Crookes,Chemical News,IV (1861)293. W.A.Miller and W.Huggins,Phil.Trans.,CLIV (1864)413-35,437-44, and Chemical News,IX(1864)277. G.Wilson,Electricity and the Electric Telegraph, together with the Chemistry of the Stars,new edn(1859); the latter essay. See also chap.9,note 17.

16. Lord Rosse himself was very cautious about the interpretation of his data; Scientific Papers,ed. C.Parsons(1926); see above, chapter 4. J.N.Lockyer, Nature,XIX(1878)197-201,225-30; W.H.Brock,'Lockyer and the Chemists; the first dissociation hypothesis',Ambix, XVI(1969)81-99. Lockyer's papers on the Sun are collect- ed in his Solar Physics(1874) in which his dislike of Huggins is apparent. On Lockyer, see A.J.Meadows, Science and Controversy(1972).

17. On Brodie and his chemical calculus, see W.H. Brock(ed.),The Atomic Debates(Leicester,1967). B.C. Brodie,Phil.Trans.,CLX(1850)759-804; see especially pp.759,781,804; for his views on the union of theory and experiment, see Proceedings of the Ashmolean Society,II(1843-52)340-4. On Ampere, see S.Mauskopf, 'The Atomic Structural Theory of Ampere and Gaudin', Isis,LX(1969)61-74.

18. B.C.Brodie,Phil.Trans.,CLIII(1863-4)407; Chemical News, IX(1864)258; Phil.Trans.,CXLIX(1859)249; Phil. Trans.,CLXII(1872)435-84; CLXIV(1874)83-103; the last two are the famous papers on ozone. On chemical symbols: W.Whewell,Phil.Mag.,IV(3rd series,1834)9; R.Phillips, same journal,p.246; J.Prideaux,Phil.Mag.,X(2nd series, 1831)104; IV(3rd series,1834)41,464; J.Daubeny,Report of the British Association(1836)xxix; J.F.W.Herschel, Report of the British Association(1858)41; W.Odling, Report of the British Association(1864)21. See also

D.M.Knight,Atoms and Elements,2nd edn(1970)ch.6.

19. B.C.Brodie,Proc.Roy.Soc.XV(1866-7)136-9; Phil. Trans.,CLVI(1866)781-859; CLXVII(1877)35-116. A report on the reading of an abstract of the first part to the Royal Society appears in Chemical News, XV(1867)269, and the text of 'Ideal Chemistry' on pp.295-302; this was the lecture delivered before the Chemical Society a revised version of which was published later: Ideal Chemistry(1880). See also J.Harris and W.H.Brock,'Alexander William Williamson', Annals of Science,XXXI(1974)95-130.

20. The discussion is reported in Chemical News,XV (1867)302-5; E.R.Paul, 'Alexander W.Williamson on the atomic theory',Annals of Science,XXXV(1978)17-31.

21. B.C.Brodie,Chemical News,XV(k867)301,302. On Victor Meyer and Chlorine, see Nature,XXI(1880)491-2; Chemical News,XL(1879)232,292.

22. C.A.Russell,The History of Valency(Leicester,1971). J.Anderson, Report of the British Association(1867)28; H.Roscoe,Journal of the Chemical Society,XXXIX(1881) 182-3; the Dictionary of National Biography is more brutal: 'We find no record of any work of interest between 1862 and 1880, when he died'. See the review of Ideal Chemistry in Nature,XXIII(1880-1)141; and the remarks of Crum Brown in Phil.Mag.,(4th series) XXXIV(1867)129-36. On valency theory, see O.T.Benfey (ed.) Classics in the Theory of Chemical Combination, (New York,1963).

23. N.R.Holt, 'A note on Wilhelm Ostwald's 'Energism'', Isis,LXI(1970)386-9; T.H.Levere,Affinity and Matter (Oxford,1971)ch.7. P.G.Tait declared that the Second Law of Thermodynamics contained implicitly the whole theory of chemistry:Report of the British Association (1871)5; and Crum Brown in 1874 urged the need for a union of chemistry and dynamics:Report of the British Association(1874)45 — cf his opposite number in the physics section,p.1. W.Ostwald, 'Faraday Lecture', Journal of the Chemical Society,LXXXV(1904)506-22. Letters to and from Ostwald, including correspondence with Willard Gibbs, appear in H.G.Korber(ed.), Aus dem wissenschaftlichen Briefwechsel Wilhelm Ostwalds, I(Berlin,1961).

24. J.H.Gladstone,Phil.Mag.,V(4th series,1853)313-20, and Nature,XXVIII(1883)500-3 (his Presidential Address to the chemical section of the British Association); B.Stewart,Nature,XXI(1880)25 included arguments from spectra for the complexity of the elements; Lord Rayleigh,Chemical News,XVI(1882)95-7, an extremely interesting Presidential Address to the physics

Section of the British Association on the question of accuracy and precision in the experimental determination of quantities.

9

Chemical Taxonomy

Ralph Cudworth remarked, in a different context, that
'Things are <u>Sullen</u>, and will be as they are, what ever
we Think them, or Wish them to be: and men will at
last discover their Errour, when perhaps it may be
too late. <u>Wishing</u> is no <u>Proving</u>'.[1] While all
empirical scientists would agree with him in principle,
we have seen how in their discussions of the unity of
matter chemists and physicists in the nineteenth
century refused to give up an attractive theory merely
because it was empirically falsified. The belief
that there were few — perhaps only one or two — kinds
of matter went well with what we can call the Newtonian view of the world, a view which survived from
Newton's own time down to the nineteenth century; it
received support from analogies, physical and chemical;
and it seemed to go well with a view of the world as
simple and harmonious. In two forms it was turned
into a falsifiable hypothesis; but although in the
twentieth century transmutations were observed in the
laboratory, and in the doctrine of isotopes a version
of the view that all matter is composed of hydrogen
atoms in various arrangements is today accepted,
neither the theory of Samuel Brown nor that of Prout
could be verified.

Wishing did in a sense turn out to be proving;

but the proofs came late in the day. That belief in the unity of matter was not confined to fanatics — it did not go with belief in astrology or the flat Earth — is an indication of the strength of the analogical arguments, and those from the consistency and simplicity of nature. But so far we have considered chiefly arguments from consistency with physics, with an occasional glance at physiology for the doctrine that organisms transmute. In this chapter we must turn to what through much of the nineteenth century, as in the eighteenth century, seemed the most important part of natural history — classification.[2] We have met chemists referring elements to their families or natural groups, and urging for example that substances so close in chemical properties as chlorine, bromine, and iodine could not be utterly distinct bodies. But such terms as 'natural group' had in the nineteenth century a considerable theory-loading; and when we turn to the writings of chemists engaged in classification, we find them busy in an area where rules for the cultivation of which had become highly developed.

That natural history was prestigious and popular in the era of Buffon, Banks, and Cuvier needs no stressing; but it is perhaps less recognised by historians of chemistry that natural history provided a paradigm for chemistry.[3] The object of natural historians, particularly in the period of great explorations which followed the publication of the Linnean system, was to describe and to classify. This was a period rich in discoveries for natural historians and for chemists alike; new species and new elements were brought to light, and creatures — and substances — which had been supposed to belong to one group were shown upon closer examination to belong to distinct

groups. Those who objected to the Newtonian or
Daltonian model for chemistry as hypothetical and unverifiable could derive comfort from the way in which
sound natural historians similarly deplored the
extravagancies of the evolutionary doctrines of
Erasmus Darwin and of Lamarck. Neither in chemistry
nor in natural history, it could be argued, was there
any need to explain, or to invoke hypothetical
mechanisms; in both sciences, the object was to
describe unambiguously — in a technical and theory-
free language — and to classify conveniently.

In the opening years of the nineteenth century,
this last objective was made harder to attain with
the demand that the system of classification be not
only convenient, but also natural. The Linnean system
had enabled naturalists to place plants and animals
in their classes, but it was confessedly artificial.
Members of the same class often differed in many important characteristics; but if they agreed in the
characters which formed the basis of classification,
then they had to go into the same pigeon-hole whatever misgivings the naturalist guided by his intuition might feel about it. Aristotle had urged that
there were in nature genuine groups of things, and
that the taxonomist should consider multiple criteria;
placing an animal or plant in its class if it agreed
with members of the class in a majority of characteristics, or at least of important characteristics.[4]
We saw Davy doing this when after deliberation he
placed potassium among the metals; it is a process
which requires experience and judgement. It also
entails the belief that there are genuine natural
kinds, and this many philosophers at the present day
are reluctant to allow: to the nominalist it is no
more scientific to put a goose among the Anatidae

247

than to put it among the Edibilia; but scientists of
our period did not espouse this kind of nominalism.

We are now ready to approach the schemes of
chemical classification produced during the nineteenth
century; schemes which indicated that there were
natural kinds among chemical elements and compounds,
having regard first to chemical properties and then
to atomic or equivalent weights. The relative weight-
ing to be given to these qualitative and quantitative
aspects of chemical substances was something which
troubled the chemical taxonomist, and made his job as
hard as that of the natural historian although there
are incomparably fewer elements than there are
animals or plants.

The eighteenth century had, in the Tables of
Affinity produced by E.F.Geoffroy and others, including most notably Tobern Bergman, a kind of classification.[5] It went with the hope for a dynamical chemistry.
When tables had been drawn up indicating which
elements would displace others from their compounds,
then it might prove possible to quantify this data
and make chemistry a science where exact predictions
were possible. As we have seen, this hope was not to
be realised, and weights were the quantities which the
taxonomist had to take into account. Displacement
reactions depend, as Berthollet was to show, upon con-
centrations and solubilities and do not give unambiguous measures of affinity.

Nevertheless, we do find nineteenth-century
examples of affinity tables, notably those published
in the Philosophical Transactions by Thomas Young,
whom we do not usually think of as a chemist.[6] As a
result of his experiments, he believed 'that nearly
all the phenomena of the mutual actions of a hundred
different salts may be correctly represented by a

hundred numbers, while, in the usual manner of relating every case as a different experiment above two
thousand separate articles would be required'. He
had 'reason to think' that cases where the numbers
wrongly predicted the course of a reaction — because
of the presence of another substance, or for some
other reason — would be few, and could soon be
noticed and allowed for. The tables were therefore a
convenient shorthand way of summarising numerous
reactions; the numbers which he produced being
'tolerable approximations' to measures of the forces
between the substances. If the principle were accepted, 'that the attractive force, which tends to unite
any two substances, may always be represented by a
certain constant quantity', then it would be possible
to represent the facts of double-decompositions
'independently of any hypothesis, and without being
liable to the contingent necessity of any future
alteration, in order to make room for the introduction
of the affections of other substances'.

This convenient principle turned out unhappily to
be false, and Young's seems to be the last serious
attempt to quantify chemistry on a basis of elective
affinities. Whether or not Young was a good Newtonian
in his optical papers, he stands out as one in this
intrusion into chemistry; and he is typical of his age
in the final paragraph where he attempts to reform the
orthography of the science, following those 'certain
rules of etymology, which mo modern author has a right
to set aside'. His tables were set out under the
names of various acids and bases; and the problem was
that while the order of strength was much the same
under each, it was not exactly the same. Thus for
'barita'(sic), 'sulfuric' acid had a greater affinity
than oxalic; for lime, it was the other way about.

Where one might have hoped that the study of affinities would have led to a single series of chemical elements and compounds each having a certain fixed affinity value, the facts proved otherwise. The elements could not be happily organised into a 'great chain' on the basis of affinity measurements; and attempts to organise them thus by means of electrical measurements also failed in the state of electrical knowledge available in the early nineteenth century. Davy and Berzelius could not produce an unambiguous scale of electronegativities for the same reasons as Young could not produce a scale of affinities; they did not seem to be strictly constant.

Just as naturalists had to abandon the attempt to place all animals and plants on the appropriate rung of the ladder or great chain of being, so chemists had to turn to classifying in groups rather than in a continuous series. The earlier attempts at this were based entirely upon chemical analogies, and included no quantitative aspect; they are therefore very comparable to those made by natural historians. Lavoisier's system — and similar schemes are found in chemical textbooks of the early nineteenth century, such as Mrs Marcet's — was based upon oxygen as the most important agent;[7] this should not surprise us in view of Lavoisier's doctrines of combustion and acidity. Lavoisier's first class (being the protagonist of the new definition of 'element', he classified elements) included light, caloric, oxygen, azote, and hydrogen. These seem an odd kind of natural group, but all are elastic fluids, imponderable or ponderable. The next group comprises the non-metals which can be oxidised and acidified, and includes sulphur, charcoal, phosphorus, and the unknown radicals of the muriatic, fluoric, and boracic acids. Then comes the class of

metals which are 'Oxydable and Acidifiable', a class
which, reasonably enough, includes arsenic, antimony,
manganese, and molybdena, but strangely also contains
also gold and iron. The last class is the salsifiable
earths, lime, magnesia, barytes, argill, and silex;
these Davy and Berzelius were to analyse with the
electric battery.

Davy's researches on chlorine removed oxygen
from its unique position, and changed the received
view of acidity; and clearly Lavoisier's scheme of
classification would no longer do. That indefatigable
classifier A.M.Ampère, famous for his electrical work,
published in 1816 an Essay towards a Natural
Classification of Simple Bodies.[8] He wrote that a
chemical classification could be artificial, but
ideally should 'be deduced from the ensemble of the
characters of bodies which we propose to classify; and
by constantly uniting those presented by the most
numerous and essential analogies, they will be to
chemistry what the natural methods are to botany and
zoology'. The old method of ranging simple bodies
according to their degree of affinity for oxygen must
be dropped and the chemist must 'begin by assigning
to each simple body the place which it ought to occupy
in the natural order, by comparing it successively
with all the rest, and uniting it with those which
resemble it by a greater number of common characters,
and particularly by the importance of those characters'.

If chemists had had a natural system of classification, then all bases would have been put into one
class, and chemists would have heated soda and potash
with iron and got sodium and potassium twenty years
sooner; that is, a natural system would assist the
chemist in the search for true analogies. Once the
elements had been classified, then the compounds —

251

which were better arranged already — could also be
incorporated into a natural system. The old distinction between metals and non-metals seemed to have a
basis in nature, but there were the elements arsenic,
tellurium, and silicium — our silicon, this form of
name indicating that it is no longer counted as a
metal — which it was very difficult to place definite-
ly on either side of this line. The division into
combustibles and supporters of combustion also failed
to lead to a clear division, and furthermore put into
the same class bodies which displayed few analogies
with one another. Ampère's own attempts 'to establish
a natural order among simple bodies, by arranging
them in a single series, which should commence with
those whose properties presented the most complete
opposition to those of the bodies which [he] attempted
to place at the end of the series' had met with no
success.

Ampère urged that he had been guided towards a
natural arrangement not in the spirit of system,
which he disavowed, but after making many attempts to
bring out chemical analogies, and considering the
elements under every point of view. His actual system
is something of a disappointment after this acute and
interesting discussion.[9] He begins with three great
groups; the gazolytes, which can form permanent gases
with one another; the leucolytes, which are metals
fusible below 25° Wedgwood — brass melts below this
temperature, and copper just above — and whose oxides
form colourless solutions with colourless acids; and
the chroicolytes, which melt above 25° and whose
oxides form coloured solutions in colourless acids.
This classification then proceeds through an hierarchy
of genera and species; in which nitrogen is put with
oxygen and sulphur, silver is put into a different

252

class from copper and gold (which themselves are in
different genera), and osmium goes with titanium in
a different genus from the other platinum metals
(which are put with gold). The system seems, in
short, as artificial as any of its predecessors and
no more convenient; and it was not only because of
the neologisms that chemists shunned it. It is a
guide only to what properties Ampère believed
important.

Ampère's classification represented an attempt
to take over biological modes of proceeding directly
into chemistry; but by the time he was writing it was
clear to most chemists that one of the most important
properties attending each element, and hence one which
any natural system must take into account, was the
atomic or equivalent weight. Schemes of classification which ignored this could only be artificial; and
such schemes in fact developed into the tables used
in qualitative analysis where the objective is not to
achieve a natural classification but to separate the
elements methodically and rapidly. It soon became
clear that the equivalent weights of elements which
seemed close in chemical properties were often in some
simple numerical relationship. Thus J.W.Dobereiner
in 1817, and at greater length in 1829, in attempting
'to group elementary substances according to their
analogies' drew attention to triads of related
elements the atomic weights of which were either very
close — like platinum, iridium, and osmium — or formed
a series in which the atomic weight of the middle
member was the mean of those of the first and third,
as in chlorine, bromine, and iodine.[10] Max Petten-
kofer showed that these triads could be represented
as parts of series; and we saw earlier how J.B.Dumas
tried to fit the atomic weights of the elements into

a series like that which connects the equivalent
weights of the homologous series of organic radicals.

In 1854 the American J.P.Cooke, who was Professor
of Chemistry at Harvard and also wrote works on
natural theology, published an elaborate table of the
elements with an essay explaining and supporting it.[11]
This did not fall upon utterly deaf ears; it appeared
in two American journals, and was referred to by Sir
John Herschel in a Presidential Address to the
British Association; but it is in some ways a curiosity.
He had succeeded in drawing up all the elements — and
cyanogen, which comes between fluorine and chlorine —
into six series, in each of which the equivalent
weights were represented by a series, such as 8+n9 or
8+n6. To fit these series the equivalents are taken
as whole numbers or as ending in .5; but the table
does not necessarily involve Proutian speculations.
It was intended as a scheme of classification 'like
those which so greatly facilitate the acquisition of
the natural-history sciences'. Every teacher, he
wrote, must have felt the need for such a system; and
Cooke's table was designed to be exhibited in the
lecture room. Following Ampère, to whom he does not
refer, Cooke gives an extended argument for a natural
system based upon true and complete analogies; the
difficulty with most classifications previously
proposed had been, in his view, 'undoubtedly, that
they are too one-sided, based upon one set of
properties to the exclusion of others, and often on
seeming, rather than real resemblances'. This was
particularly true of the distinction between metals
and non-metals, which separated phosphorus from
arsenic and sulphur from selenium.

Cooke believed that there was no one fundamental
property upon which a chemical classification might

254

be founded; and that chemists 'must, therefore, here, as in other sciences, base our classification on general analogies'. The most important chemical property was, he declared, undoubtedly crystalline form; but as a basis for arrangement it was useless. His own classification was presented as provisional, as one which had proved its usefulness in teaching; but it had also led to the discovery of numerical relations between atomic weights, and it was for this reason that he published it. As well as the inclusion of cyanogen among the elements, the table has the peculiarity that oxygen appears in more than one series. All the relations were 'not that of a mere accidental group of numbers, but [were] connected with the most fundamental properties of the elements'. He hoped for further quantitative data, which would bring out the law of the arrangement of the elements; but although he compared his series to those of homologous radicals he did not, in the paper, advance any speculations about the complex structure of the elements.

Classifiers are perhaps rarely prone to welcome the idea that what they are classifying is unstable; and the most serious among the classifiers of the chemical elements seem to have taken it for granted that they were arranging fixed entities rather than mere radicals. In the 1860s there was agreement at last on atomic weights, which brought the possibility of arrangements involving both a sequence of numbers — the atomic weights, for equivalents do not work — and a grouping into natural families; and during the decade the Pythagorean speculations of de Chancourtois and the Periodic Tables of Odling, Newlands, Meyer and ultimately Mendeleev were published.[12] To Mendeleev would seem to be justly due the greatest credit, for

— as he remarked — he used the Table to make detailed predictions. One test, after all, of a natural classification, as of a theory, is that it should lead to the perception of true analogies and hence to predictions of this kind. Mendeleev's predictions were based upon chemical analogies, and in the famous case of tellurium and iodine he put the elements into the families where their chemical properties indicated that they should go, even though this ran against the order of atomic weights. He was not therefore open to the charge made by Marcellin Berthollet that the construction of such tables was a matter of playing about with numbers and not a serious chemical activity; a charge on which many of the pioneers of chemical classifications might be found guilty.

Mendeleev himself was opposed to the belief in the unity of matter. The Periodic Law was for him a matter of observation, which entailed no Pythagorean or Proutian consequences. The analogies between elements and radicals were imperfect, the study of spectra did not compel one towards belief in the unity of matter, and, he declared in his Faraday Lecture of 1889, 'from the failures of so many attempts at find-ing in experiment and speculation a proof of the compound character of the elements and of the existence of primordial matter, it is evident, in my opinion, that this theory must be classed amongst mere utopias'. He added that 'experiments and attempts at confirming that idea have been so numerous that it really would be instructive to have them all collected together, if only to serve as a warning against the repetition of old failures'. Laws founded on the basis of experiments were, in his opinion, like the telescope and the microscope, the means of enlarging our mental horizon. Mendeleev himself produced some curious

queries about weight and motion, and at the end of
his life published a speculative little work, A
Chemical Conception of the Ether; and his attempt
to stop chemists using his Periodic Table as a basis
for Proutian speculation was unsuccessful.[13]

Even though Mendeleev's Table did not have the
genera and species like those of Ampère and Cooke, it
did group the elements into families, at first in
horizontal lines and subsequently in vertical columns.
Other forms of it brought together horizontal families, the transition metals and the rare earth metals.
Whatever Mendeleev may have hoped, it was natural
that men of science in the period in which Darwinian
theory (almost universally accepted even if only as a
working hypothesis) was transforming biology, should
see the Periodic Table as not only a guide to true
relationships but also as an indicator of community
of descent. As the American palaeontologist A.C.
Marsh said in an address to the American Association
for the Advancement of Science in Tennessee in 1877:
'I am sure I need offer here no argument for evolution;
since to doubt evolution today is to doubt science,
and science is only another name for truth'. Van
Spronsen has described the various forms into which
the Periodic Table was cast by people who wanted to
bring out more clearly the various relationships which
it summarises; and we shall not look any longer at the
Table for its own sake, but only in its relation with
theories of the unity of matter and of inorganic
evolution.

Belief in inorganic evolution had preceded
Mendeleev's Table. Thus in 1867 Thomas Sterry Hunt
delivered at the Royal Institution a lecture on the
chemistry of the primeval Earth.[16] He was an
American geochemist; and began by referring to the

257

nebular hypothesis, rendered respectable once more
through the labours of the spectroscopists. The
paper was printed in William Crookes' journal,
Chemical News; and the body of it contains a discussion of the cooling of the Earth and the formation of
the rocks. In the primeval Earth the temperature
would have been so high as to dissociate all compounds
into their elements, which were themselves perhaps
dissociated. In a further lecture in 1882 Hunt
invoked Newton's aether, and his idea of the conversion of rare into dense matter as the source of the
carbonic acid in the atmosphere, which otherwise
would have been fixed in the rocks.

These aetherial speculations were followed up by
Thomas Carnelly, who in 1886 read a paper before the
British Association on 'The Cause of the Periodic Law
and the Nature of the Chemical Elements'.[15] His work
was intended as a generalisation of that of J.B.Dumas,
whom we saw earlier as advancing the view that the
elements were analogous to the homologous series of
organic radicals. Carnelly tried to incorporate these
analogies into Mendeleev's Table in the hope that he
could thereby explain it, for there was no real
explanation of the periodic properties of the elements
until the rise of atomic physics and in particular the
work of Moseley and Bohr. A consequence of this was
that many more elements were predicted from Tables
than were in fact discovered; and that particularly
among the rare earths, where no analogies were
available, the Table gave no clue as to how many
elements there might be. Carnelly's tables indicated
the parallels between elements and radicals; and therefore, he concluded, they were 'analogous in both form
and function'. But whereas the hydrocarbon radicals
were composed of carbon and hydrogen only, the

elements were composed of at least two simple
elements, and were probably compounds of carbon and
ether of atomic weight -2. Hence there were but
three elements, hydrogen, carbon, and ether.
Carnelly added that this bold speculation would need
'ample confirmation — and no doubt some modification'
before it was generally accepted.

Carnelly's paper was also published in Crookes'
Chemical News; this was the chief vehicle in English
for theoretical or speculative papers, but Crookes
took a particular interest in speculations about the
nature of the elements as we saw in discussing the
theory of the fourth state of matter. A little later
in 1886 the same journal published Emerson Reynolds'
'Method of Illustrating the Periodic Law'.[16]
Carnelly was Professor of Chemistry at St.Andrew's,
and Reynolds occupied the same position in Dublin.
He referred to the 'numerous and important papers' of
Carnelly, which had made known the details of the
Periodic Table. Reynolds' arrangement put the
elements on a zig-zag which crossed the centre line
between hydrogen and lithium, fluorine and sodium,
chlorine and potassium, nickel and copper, and so on.
He compared the elements to knots on a string arranged in this zig-zag; they can be imagined to form a
string or chaim — the physicist George Fitzgerald,
ever fertile in helpful ideas, suggested a chain
hanging from the roof to the floor and fixed at both
ends — whose parts are in unequal tension and vibrating; they could then be viewed as composing a kind of
vibrating system. One cannot but admire the
Victorians' capacity to apply mechanical models like
this; but Reynolds' model did help him to make certain
predictions. Thus he inferred that hydrogen did not
stand by itself as it seemed to do in the ordinary

representation of the Table; on its side of the centre line there was room for seven elements beneath it, one of which might be the helium recently announced by Lockyer as existing in the Sun although it had not yet been isolated upon Earth. There were no elements where the zig-zag crossed the centre line the first three times, but three of the higher nodal points were occupied by iron, cobalt, and nickel, and by the two groups of platinum metals. These, Reynolds remarked, had 'often been regarded as probable modifications of a single form of matter'; but he did not use his model in any account of the genesis of the elements.

This was done by Crookes himself.[17] He constructed a three-dimensional model of the Periodic System in which instead of being on a zig-zag the elements appear on a figure-of-eight spiral; this model may be seen at the Science Museum at South Kensington, London, and interested William Ramsay when he was perplexed to fit the family of inert gases into the Periodic System. In the late summer of 1886 he, as President, addressed the Chemical Section of the British Association, taking as his subject the problem of the elements. He quoted the usual authorities, Faraday, Lockyer, and Brodie for the belief that the elements were complex, and mentioned also Herbert Spencer, Thomas Graham and G.G.Stokes. Turning to the Periodic Table, he remarked that 'the array of the elements cannot fail to remind us of the general aspect of the organic world'. The distribution of elements even displayed some parallels with what is observed among animals, the curious group of rare earth metals, confined apparently to Sweden, being perhaps analogous to the monotremes of Australia. Crookes admitted that there was no direct

evidence that the Table represented an evolutionary
sequence, and that the parallels with the biological
realm were incomplete; in particular, there was no
fossil record of extinct elements. He referred to
the view that Lockyer's helium might be the one true
element, with an atomic weight a fraction that of
hydrogen; it was of course convenient for Proutians
of this new school that at this date there was no way
of measuring the atomic weight of this new candidate
for the role of 'protyle'.

But more important for Crookes were the problems
associated with the rare earth metals, upon the
separation of which he had been engaged. Mendeleev's
Table gave no clue as to how many of these metals
there might be, because there was no analogous complete
series. Crookes became inclined, as he declared to
the British Association and later in a Presidential
Address to the Chemical Society of London, to reject
the view that these metals represented a family of
distinct bodies. Rather he seemed to find a continuum.
Evidence from spectroscopy and from chemical reactions
supported this view; we do now believe these metals to
be a group of distinct elements, but they are extremely difficult to separate from one another and with
only Lavoisier's definition of 'element' to guide one
it would be difficult to know how to describe them.

Crookes now turned to Reynolds' version of the
Periodic Table, which he had published two months
before. But he represented it as the swings of a
pendulum, which starting from hydrogen — or really
from protyle — swung out to carbon, back to silicon,
back again to titanium, and so on, the swings decreas-
ing in amplitude each time. The lightest elements
were thus the first born; rapid cooling yielded like
elements, and where the cooling had been extremely

rapid then, as among the rare earths, elements not properly differentiated had resulted. Each element, moreover, was probably not composed of particles having exactly the same mass, but rather clustering about a mean value. Among his arguments for evolution he referred to the 'weighty arguments' of Carnelly, and to the writings of Lockyer; for the problem was that this process of inorganic evolution was probably no longer going on upon the Earth, and could only be observed — if at all — in the heavens. Before it was sensible to seek to find sub-atomic particles or to investigate atomic structure, it was necessary to believe that the chemical elements were not true simple bodies; but before any explanation could be given of Crookes evolutionary pendulum it was necessary to have atomic physics. To the various rare earth metals Crookes was unwilling to extend the term 'element'; he suggested that such 'doubtful bodies' be described as meta-elements, and he declared that

> if the study of the rare earths leads
> us to clearer views on the nature of
> the elements, neither my colleagues
> nor myself will, I am sure, regret
> the months spent in tedious and
> apparently wearisome fractionations
> ... we can only work on, 'unresting,
> unhasting', trusting that in the end
> our work will throw some white
> light upon this deeply interesting
> department of chemical physics.[19]

In 1891 in his famous paper on Electricity in Transitu — which was his Presidential Address to the Institution of Electrical Engineers — Crookes used the same diagram derived from Reynolds' depiction of the Periodic Table. After describing brilliant experiments on the cathode rays, he moved at the end to 'show how intimately chemistry and electricity interlock' in describing how electrical polarisation was involved in a consistent explanation of the genesis of the elements. Hoping that he might be excused for digressing into this development of theoretical chemistry, he remarked that it was now 'generally acknowledged' that there were ranks in the hierarchy of the elements; that 'besides the well-defined groups of chemical elements, there are underlying sub-groups' to which the name 'meta-elements' has been given.[20]

Crookes believed that two forms of energy were required, 'one operating uniformly in accordance with a continuous fall of temperature' which moves the process steadily down the Table towards elements of greater atomic weight, 'and the other having periodic cycles of ebb and swell, and intimately connected with the force of electricity', to produce the zig-zags back and forth with the production of electropositive and electronegative elements: 'The centre of this creative force in its journey through space scattered seeds or sub-atoms that ultimately coalesced into the groupings known as chemical elements.' Crookes used ideas he had developed in discussing the formation of dark spaces and striations in the discharge tube to describe how these 'new-born particles' would form groups in different parts of space as 'the faster moving ones would still overtake the laggards, [and] the slower would obstruct the quicker'.

Within each group, those constituent particles 'whose form of energy governing atomic weight was not in accord with the mean rate of the bulk of the components of that group, would work to the outside and be thrown off to find other groups with which they were more in harmony'. Thus stable chemical elements would result, each having a definite atomic weight, 'definite on account of its being the average

weight of an enormous number of sub-atoms or meta-
elements, each very near to the mean'. Thus the
atom of mercury is to be envisaged as composed of
'an enormous number of sub-atoms, each of which may
vary slightly round the mean number 200 as a centre'.
The sub-atoms are not sub-atomic particles; they are
individual atoms, while the atom of mercury of atomic
weight 200 would seem to be a statistical average,
applying only to large numbers.

To the old question of why, if the elements have
been evolved we cannot transmute them, Crookes had an
answer ready; 'the question is as futile as the cavil
that in the organic world we never see a horse meta-
morphosed into a cow'. Before copper or another metal
could be transformed into gold, 'it would have to be
carried back to a simpler and more primitive state of
matter, and then, so to speak, shunted on to the track
which leads to gold'.

As the pendulum swung towards the centre line,
electropositive elements were generated, and as it
swung away electronegative. An atom in the mean
position would be neutral; sub-atoms either side of
the line would have a positive or negative charge.
To Crookes this account of the genesis of the elements
was not 'a mere hypothesis, but may take the rank of
a theory. It has been experimentally verified as far
as possible with so baffling an enigma'. The
researches of Crookes and his associates on yttria
had shown that in matter which fulfilled all the
criteria of elementary nature, there were minute
shades of difference which had admitted of 'resolution
into meta-elements, having exactly the properties
required by theory'. Crookes added that he had

> struggled with problems which must
> be conquered before we can arrive

at exact conclusions — conclusions
which, so far as inorganic nature
is concerned, can only be reached
by the harmonious interfusion —
not confusion — of our present twin
sciences, electricity and chemistry.
Of this interfusion I have just
endeavoured to give you a foretaste.
In elaborating the higher physics,
the study of electrical phenomena
must take a large, perhaps the
largest, share. We have invaded
regions once unknown, but a
formidable amount of hard work
remains to be completed. As we
proceed we may look to electricity
not only to aid, as it already does,
our sense of hearing, but to sharpen
and develop other powers of perception.

The rhetoric reminds us of Priestley and of Davy,
inaugurating the interfusion of electricity and
chemistry a century earlier; and like them Crookes
was not afraid of being speculative or of using his
researches as a basis for conclusions in widely-
separated regions. His theory of the evolution of
the elements, with its curious scaffolding of pendulum
and unknown forms of energy, and its urgent and grand-
iloquent manner of presentation, reminds us perhaps
more of Vestiges than of the Origin of Species; but it
would be unfair to demand of a pioneer that he should
always coolly present a fully-articulated theory.
Crookes' mechanism was new — if we stretch the term
to call it a mechanism — and so was his insistence
that the Periodic Table must be a guide to the origin
and distribution of the elements as the natural systems
were in the biological realm. His insistence that
electricity was the key to the mystery of the nature
of the chemical elements was not new, though he gave
it new forms; and some of his other points echo in
different language those we have seen earlier chemists

making. Thus the strong similarities between the various alkali metals and between the various platinum metals had inclined many chemists to regard them not as distinct species; the even-stronger similarities between the Rare Earth elements inclined Crookes to take a similar view of them, which he elaborated into his theory of meta-elements.

Within a decade of Crookes' address to the electrical engineers, X-rays had been discovered by W.C.Röntgen and J.J.Thomson had published his paper determining e/m for the corpuscles of which he showed the cathode rays to consist.[21] Thomson supposed that these corpuscles might be that prime matter for which chemists had long sought; and it was not long afterwards, in 1903, that the first transmutation was observed in the laboratory. Ramsay had used Crookes' version of the Periodic Table in his attempts to predict the properties of the congeners of argon; he supposed that they might form a cluster about the central line like the platinum metals, and therefore form a triad having approximately identical atomic weights. Early in 1895 Ramsay had succeeded in isolating helium which had been occluded in a mineral, and Crookes had confirmed the identification of its spectrum with that of the solar element. Within the next few years Ramsay isolated the rest of the inert gas family, whose atomic weights turned out to be different from the results predicted from the zig-zag Table. With the discovery of radioactivity and the preparation of radium came a particular interest in the inert emanation, which we call radon, which was given off from radium; and when Rutherford's collaborator Frederick Soddy came to Ramsay's laboratory they investigated this gas and followed its radioactive decay into helium.

The atomic models of J.J.Thomson and of
Rutherford showed the atoms of the elements to be
complex structures,[22] and the discovery in 1913 by
Theodore Richards of Harvard that lead found in the
neighbourhood of uranium had a lower atomic weight
than that found elsewhere might be thought to confirm
the speculations of Crookes on the meta-elements.
But so long as Lavoisier's negative-empirical definition of 'element' held sway, it was impossible to know
how many rare earth metals there were, or whether all
were meta-elements. H.G.J.Moseley, by his studies on
X-ray spectra of the elements, and Rutherford by his
scattering experiments with α- rays, established a new
definition of 'element', that all the atoms of an
element had the same positive charge on the nucleus.
Chemistry and electricity were now firmly married;
atomic weight determinations became less interesting
and fundamental; and what had been the transcendental
part of chemistry had become at least indirectly
verifiable.

A-propos of a discussion of the luminosity of
the sea, J.Macartney remarked in 1810 that the

> extraordinary advances lately made
> by Mr. DAVY, on the decomposition
> of substances, that were hitherto
> looked upon as elementary, give us
> reason to hope, that future invest-
> igations may unfold views of the
> material world, of which we can at
> present have only an indistinct
> conception; that new modes of
> analysis may enable us to see things,
> not 'through a glass darkly', but
> more nearly as they are; and that
> the boundaries of physical and meta-
> physical science, now far asunder,[23]
> may be made to approach each other.

These words were echoed by Crookes; but the various
researches and speculations on the nature of the

elements did not transform chemistry into a branch
of metaphysics or vice-versa.

One point which does emerge from all these
writings is that a fairly general acceptance, based
upon a range of more or less firm analogies, kept
alive the belief that the elements were complex
despite a number of apparent falsifications.
Chemistry seems in some ways a science for the hard-
headed; but during the nineteenth century adherence
to this metaphysical doctrine of the simplicity of
nature led to the relatively frequent publication of
papers on the nature of the elements in respectable
journals and by respectable scientists, and in the
twentieth to the surprisingly rapid and easy acceptance of the fact of spontaneous transmutation as sub-
atomic chemical change.

NOTES

1. R.Cudworth,The True Intellectual System of the
Universe(1678)p.888. On the general topic of this
chapter, see J.W.Van Spronsen,The Periodic System of
Chemical Elements:a history of the first hundred
years(1969), and the unpublished thesis on Mendeleev
by J.R.Smith(London,Chelsea College,1976).

2. W.Blunt,The Compleat Naturalist:a life of Linnaeus
(1971); see especially the appendix by W.T.Stearn on
the Linnean System; and F.A.Stafleu,Linnaeus and the
Linneans(Utrecht,1971); see also A.M.Lysaght,Joseph
Banks in Newfoundland and Labrador,1766(1971); this
book gives a useful picture of Banks as a thoroughly
competent natural historian and an energetic promoter
of experiment and exploration, as does H.B.Carter,
'Sir Joseph Banks...',Bulletin of the B.M.(Natural
History),Historical Series,IV(1974)281-385.

3. On Buffon, see A.Thackray,Atoms and Powers
(Cambridge,Mass.,1970)pp.155-60; on Cuvier,W.Coleman,
Georges Cuvier,Zoologist(Cambridge,Mass.,1964)p.34,
and M.P.Crosland,The Society of Arcueil(1967)pp.44-5,
222-3,and Cuvier's own Rapport Historique(Paris,1810)
pp.56-145. Lamarck also interested himself in chemistry;
see C.C.Gillispie,The Edge of Objectivity(Princeton,

N.J.,1960)pp.273-6. M.P.Crosland,Historical Studies in the Language of Chemistry(1962)part 3, describes the introduction of systematic nomenclature into chemistry through Bergman and ultimately from Linnaeus. See also D.M.Knight,'Science and Profession-alism in England,1770-1830',Proceedings of XIV International Congress(Tokyo,1974)I,53-67.

4. Aristotle,de Partibus Animalium,bk.I,chp.2-4; H.Davy,Phil.Trans.,XCVIII(1808)32; A.E.Gunther, A Century of Zoology at the British Museum,1815-1914 (1975); D.M.Knight, Zoological Illustration(1977).

5. A.M.Duncan,'The Functions of Affinity Tables and Lavoisier's List of Elements',Ambix,XVII(1970)28-42, and his introduction to the reprint(1970) of T. Bergman,A Dissertation on Elective Attractions,tr. T.Beddoes(1785); see also W.A.Smeaton, 'Schwediauer, Bentham and Beddoes:Translators of Bergman and Scheele' Annals of Science,XXIV(1968)259-73. C.L.Berthollet, Chemical Statics,tr.B.Lambert(1804); F.Szadabvary, History of Analytical Chemistry,tr.G.Svehla(Oxford, 1966)ch.12; there is an interesting account of chemical dynamics in R.Kane,Elements of Chemistry (Dublin,1840)pp.271ff,296ff.

6. T.Young,Phil.Trans.,XCIX(1809)148-60; see especially pp.148,149-50,154. For electrochemical series, see I.Newton,Opticks(New York,1952)pp.380-2 for the first steps; H.Davy,Phil.Trans.,XCVIII(1808) 364.

7. A.L.Lavoisier,Elements of Chemistry,tr.R.Kerr (Edinburgh,1790)pp.175-8; C.E.Perrin, 'Lavoisier's Table of the Elements',Ambix,XX(1973)95-105; cf T. Beddoes,Contributions to Physical and Medical Knowledge(Bristol,1799)pp.211-30, and his essay on the Chain of Beings (of 1785) in J.E.Stock,Memoirs of the Life of Thomas Beddoes(1811)appendix 2; and W.Prout,Bridgewater Treatise,2nd edn(1834)p.108; J.Marcet,Conversations on Chemistry(1806); D.R.Oldroyd, 'The concept of Earth in the 17th and 18th centuries', Annals of Science,XXXI(1974)269-306.

8. A.M.Ampère,Phil.Mag.,XLVII(1816)438-46; especially pp.438,439,444; cf his Essai sur la philosophie des sciences(Paris,1834) which represents an attempt to classify the sciences in a natural order; a task attempted at much the same time as Hegel and by Whewell; see Hegel's Philosophy of Nature,tr.M.Petry (1971) and Whewell's Philosophy of the Inductive Sciences,2nd edn.(1847)II,pp.113-8. See also K.R. and D.L.Gardiner,'André-Marie Ampere and his English Acquaintances',British Journal for the History of

Science,II(1965)235-45.

9. His system is set out by Thomas Thomson in Annals of Philosophy,IX(1817)8-10; a full translation of the actual system seems never to have been made. On Josiah Wedgwood's pyrometer and scale of temperature, see Phil.Trans.,LXXII(1782)305-26; LXXIV(1784)358-84; LXXVI(1786)390-408; in the Science Museum, South Kensington,London, there are specimens of the clay balls and grooves used in this system of measuring high temperatures; see R.E.Schofield,The Lunar Society of Birmingham(1963)pp.170-1, 262-7,and J.A.Chaldecott, 'Josiah Wedgwood(1730-95) — Scientist',British Journal for the History of Science,VIII(1975)1-16.

10. J.W.Dobereiner,in H.M.Leicester and H.S.Klickstein Sourcebook in Chemistry,1400-1900(New York,1952)pp. 268-72; J.B.Dumas,Annales de Chimie et de Physique, LV(1859)129-210. See the papers by S.C.Kapoor,'Dumas and Organic Classification',Ambix,XVI(1969)1-65; 'The Origins of Laurent's Organic Classification',Isis,LX (1969)477-527.

11. J.P.Cooke,Silliman's American Journal of Science, new series,XVII(1854)387-407; especially pp.387,389, 395,396-7,403. The paper also appeared in Memoirs of the American Academy of Sciences,V(1855)235-56, and was referred to by Herschel in Report of the British Association(1858)41. Cooke taught Theodore Richards, and set him on a course of determining atomic weights as a test of 'Proutian' views; he became one of the greatest chemical analysts of his day: see note 22.

12. J.W.Van Spronsen,The Periodic System of Chemical Elements:a history of the first hundred years(1969) is unwilling to assign especial credit to any one of these classifiers, in a spirit of detachment which is curious but perhaps admirable. M.A.E.B.de Chancourt-ois,Nature,XLVI(1889)186-8; J.A.R.Newlands,Chemical News,II(1863)70-2; X(1864)59-60; The Periodic Law (1884); W.Odling,Quarterly Journal of Science,I(1864) 642-8; J.Lothar Meyer,Annalen der Chemie,Supplement-band,VII(1870)354-64; D.I.Mendeleev,Zeitschrift fur Chemie,XII(1869)405-6; English versions of these last two may be found in H.M.Leicester and H.S.Klickstein, Sourcebook in Chemistry,1400-1900,pp.434-44. D.C. Rawson,'Mendeleev and the Periodic Law',Annals of Science,XXXI(1974)181-204;Mendeleev had a long paper published in Chemical News,XL(1879)231-2,243-4,255-6, 267-8.279-80,291-2.303-4; XLI(1880)2-3,27-8,39-40, 49-50,61-2,71-2,83-4,93-4,106-8,113-14,125-6. Except for Meyer's, all these (and many other papers referred to in this chapter) are reprinted in facsimile in

D.M.Knight(ed.),Classical Scientific Papers,Chemistry:
2nd Series(1970).

13. D.I.Mendeleev,Journal of the Chemical Society,
LV(1889)634-56, especially pp.647,648; An Attempt
towards a Chemical Conception of the Ether,tr.G.
Kamensky(1904). O.T.Marsh,Introduction and Succession
of Vertebrate Life in America(New Haven,1877)3.

14. T.S.Hunt,Chemical News,XV(1867)315-7; 'Celestial
Chemistry from the time of Newton',Chemical News,XLV
74-682-3; see above, chapter 4.

15. T.Carnelly,Chemical News,LIII(1886)157-9,169-72,
183-6,197-200; especially pp.169,184-5197,200.
S.Rozental(ed.), Niels Bohr(Amsterdam,1967).

16. J.E.Reynolds,Chemical News,LIV(1886)1-4.

17. W.Crookes,Report of the British Association(1886)
558-76; see especially pp.561,562. R.K.De Kosky,
'Spectroscopy and the Elements in the late nineteenth
century: the work of Sir William Crookes', British
Journal for the History of Science,VI(1973)400-23;
and M.Sutton,'Sir John Herschel and the Development
of Spectroscopy in Britain',B.J.H.S.VII(1974)42-60.

18. W.Crookes,Chemical News,LX(1889)27-30,39-41,51-3,
63-6.

19. W.Crookes,Report of the British Association(1886)
565-76. See p.575 for his suggestion that the negative
particles of Carnelly might be identified with
Helmholtz's particles of electricity; Chemical News,LX
(1889)66.

20. W.Crookes,Chemical News,LXIII(1891)53-6,68-70,
77-80,89-93,98-100,112-4; especially pp.113,114; for
the explanation of striations, p.55.

21. W.R.Nitske,The Life of William Conrad Röntgen
(Tucson,Arizona,1971); J.J.Thomson,Phil.Mag.,XLIV
(5th series,1897)293-316; W.Ramsay,'An undiscovered
gas',Nature,LVI(1897)378-82; W.Ramsay and F.Soddy,
Chemical News,LXXXVIII(1903)100-1; E.Rutherford and
F.Soddy,Phil.Mag.,IV(6th series,1902)370-96. M.Travers,
A Life of Sir William Ramsay(1956); R.H.Steuwer,
'William H.Bragg's Corpuscular Theory of X-rays and
γ-rays',British Journal for the History of Science,V
(1971)258-81; F.Soddy,Radioactivity and Atomic Theory,
ed.T.J.Trenn(1975).

22. A.Romer(ed.),The Discovery of Radioactivity and
Transmutation(New York,1964) and S.Wright(ed.),
Classical Scientific Papers,Physics,(1964) reprint
(the latter in facsimile), useful collection of papers
on the rise of atomic and nuclear physics. See also

S.Rozental(ed.),Niels Bohr(Amsterdam,1967); and T.J.
Trenn, 'The Justification of Transmutation',Ambix,
XXI(1974)53-77. See also the papers by P.M.Heimann,
'Rutherford, Nagoaka, and the nuclear atom',Annals of
Science,XXIII(1967)299-303; 'Moseley and Celtium:the
search for a missing element',Annals of Science,XXIII
(1967)249-60; 'Moseley's Interpretation of X-ray
Spectra',Centaurus,XII(1968)261-74; 'Maxwell, Hertz,
and the Nature of Electricity',Isis,LXII(1971)149-57.
See J.J.Thomson,Recollections and Reflections(1936);
and the obituary of Richards in Proc.Roy.Soc.,A,
CXXXI(1928)xxix. The papers of Rutherford and Moseley
appear in Wright's compilation. See J.L.Heilbron,
H.G.J.Moseley(Berkeley,1974).

23. J.Macartney,Phil.Trans.,C(1810)288.

Select Bibliography

References to articles have been given in the foot-
notes at appropriate points in the text; this
bibliography will contain only books, particularly
those of especial value in this study and general
works giving other views of chemistry during the late
eighteenth and the nineteenth century.

Primary sources

Abernethy,J. Surgical and Physiological Essays,1793

Babbage,C. Reflections on the Decline of Science in
England,1830;
Ninth Bridgewater Treatise,1837;
Passages from the Life of a Philosopher,1864

Barrow,I. The Usefulness of Mathematical Learning,
J.Kirby(tr.)1734

Beddoes,T.(ed.),Contributions to Physical and Medical
Knowledge,principally from the West of England,
Bristol 1799

Beddoes,T. and Watt J. Considerations on the Medicinal
Uses of the Factitious Airs,Bristol 1795;
Memoirs of the Life of Thomas Beddoes by J.E.Stock,
Bristol 1811

Bergman,T. A Dissertation on Elective Attractions,
T.Beddoes(tr.),1785

Berthollet,C.L. Essay on Chemical Statics,B.Lambert
(tr.),2 vols.1804

Berzelius,J.J. Essai sur la Theorie des Proportions
Chimique,Paris 1819, reprint,intr.C.A.Russell,1972;
Traite de Chimie,A.L.Jourdain and M.Esslinger(tr.),
8 vols.Paris 1829-33

273

Black,J. Lectures on Chemistry,J.Robison(ed.),2 vols.
Edinburgh 1803;
Notes from Dr Black's Lectures on Chemistry,1767-8,
by T.Cochrane, D.McKie(ed.),Wilmslow,Cheshire 1966.

Birch,T.(ed.), The History of the Royal Society of
London,4 vols.1756-7;
The Works of the honourable Robert Boyle,6 vols.1772

Boscovich,R.J. A Theory of Natural Philosophy,J.M.
Child(tr.),Chicago 1922;new edn.,translation only,
Cambridge,Mass.1966.

Brande,W.T. A Manual of Chemistry,1819;4th edn.1836

Brodie,B.C. Ideal Chemistry,1880

Brown,S. Lectures on the Atomic Theory, and Essays
Scientific and Literary,2 vols.Edinburgh 1858

The Electrical Researches of the honourable Henry
Cavendish, J.C.Maxwell(ed.),Cambridge 1879

Chambers,R. Vestiges of the Natural History of
Creation,5th edn.1856

Chaptal,J.A. Elements of Chemistry,W.Nicholson(tr.),
2nd edn.,3 vols.1795

Chenevix,R. Remarks upon Chemical Nomenclature,1802.

Coleridge,S.T. Aids to Reflection,T.Fenby(ed.),
Liverpool 1874;
Hints towards the Formation of a More Comprehensive
Theory of Life,S.B.Watson(ed.)1848; (critical
editions of these will appear in the Collected Works);
The Friend,B.E.Rooke(ed.),2 vols.1969;
Notebooks,K.Coburn(ed.),1957-

Comte,A. The Positive Philosophy,H.Martineau(tr.)
2 vols.1853

Dalton,J. A New System of Chemical Philosophy,
Manchester,pt.1,1808, pt.2,1810, pt.1 of vol.2,1827;
Memoirs of John Dalton,W.Henry,1854;
A New View of the Origin of Dalton's Atomic Theory,
by H.E.Roscoe and A.Harden,1896, reprint itr. by
A.Thackray,1970.

Daniell,J.F. Introduction to the Study of Chemical
Philosophy,1839.

Darwin,E. The Botanic Garden,1791.

Daubeny,C. Introduction to the Atomic Theory,Oxford
1831; 2nd edn.1850;
Miscellanies, 2 vols.1867.

Davy,H. Researches Chemical and Philosophical chiefly
concerning Nitrous Oxide,1800;

Syllabuses,1802 and 1804;
Elements of Chemical Philosophy,pt.1, vol.1,1812;
Consolations in Travel, or the last days of a
philosopher,1830;
The Collected Works of Sir Humphry Davy, J.Davy(ed.)
9 vols.1839-40;
The Life of Humphry Davy,by J.A.Paris,2 vols.1831;
Memoirs of the Life of Sir Humphry Davy, by J.Davy,
2 vols.1836;
Fragmentary Remains...of Sir Humphry Davy,by J.Davy,
1858

Donovan,M. Essay on Galvanism,Dublin 1816;
Chemistry,4th edn.1839

Dumas,J.B. Traité de Chimie applique aux arts,8 vols.
Paris 1828;
Leçons sur la Philosophie Chimique,M.Bineau(ed.),
Paris 1837

Edinburgh Essays,Edinburgh 1857

Essays by a Society of Gentlemen at Exeter,1796

Faraday,M. Chemical Manipulation,1827;
Experimental Researches in Electricity,3 vols.1839-
55;
A Course of Six Lectures on the Non-metallic Elements,
J.Scoffern (ed.),1853;
Experimental Researches in Chemistry and Physics,1859;
Diary,T.Martin(ed.),8 vols.1932-6;
The Life and Letters of Faraday,by H.Bence Jones,
2 vols.1870;
Selected Correspondence of Faraday,L.P.Williams(ed.),
2 vols.1971

Fourcroy,A.F. A General System of Chemical Knowledge,
W.Nicholson(tr.),11 vols.1804

Fownes,G. A Manual of Elementary Chemistry,1844; 6th
edn.1856; 11th edn.1873

Garnett,T. Outlines of a Course of Lectures on
Natural and Experimental Philosophy,1801;
Outlines of a Course of Lectures on Chemistry,1801

Good,J.M. The Book of Nature,3 vols.1826; New York
1833

Graham,T. Elements of Chemistry,2nd edn.2 vols.1850-8;
The Life and Works of Thomas Graham,by R.A.Smith,
Glasgow 1884

Gren,F.C. Principles of Modern Chemistry,2 vols.1800

Griffiths,T. Chemistry of the Four Ancient Elements,
1842

Hales,S. Vegetable Statics, 1727

Hartley,D. Observations on Man, 2 vols.1849

Heberden,W. Commentaries on the History and Cure of Diseases,1802

Henry,W. The Elements of Experimental Chemistry,7th edn. 2 vols.1815

Herapath,J. Mathematical Physics, 2 vols.1847

Heron,R. Elements of Chemistry, 1800

Herschel,J.F.W. Preliminary Discourse on the Study of Natural Philosophy,1830; Essays from the Edinburgh and Quarterly Reviews,1857

Hutton,C. A Mathematical and Philosophical Dictionary, 2 vols.1795-6

Johnston,J.F.W. The Chemistry of Common Life, 2 vols. Edinburgh 1855

Joule,J.P. Scientific Papers, 2 vols.1884-7

Kane,R. Elements of Chemistry,Dublin 1840; 2nd edn. Dublin 1849

Kirby,W. Bridgewater Treatise on the power,wisdom,and goodness of God as manifested in the Creation of Animals, T.R.Jones(ed.),2 vols.1853

Laplace,P.S. The System of the World, J.Pond(tr.) 2 vols.1809

Laurent,A. Chemical Method, W.Odling(tr.) 1855

Lavoisier,A.L. Essays,Physical and Chemical,T.Henry (tr.),1776; reprint, intr.F.Greenaway,1970; Elements of Chemistry,R.Kerr(tr.),Edinburgh 1790

Liebig,J.von, Animal Chemistry,1842,W.Gregory(tr.), intr.F.L.Holmes, reprint,New York 1964; Familiar Letters on Chemistry,1843

The Leibniz-Clarke Correspondence,H.G.Alexander(ed.), Manchester 1956

Lockyer,J.N. Solar Physics,1874; Chemistry of the Sun,1887

Low,D. An Inquiry into the Nature of the Simple Bodies of Chemistry,1844; 2nd edn.1848; 3rd edn.1856

Macquer,P.J. Elements of the Theory and Practice of Chemistry, A.Reid(tr.),3rd edn.,2 vols.1775

Marcet,J. Conversations on Chemistry,2 vols.1806; 16th edn.1853

The Scientific Papers of James Clerk Maxwell,

W.D.Niven(ed.),2 vols.Cambridge 1890

Murray,J. Elements of Chemistry,2 vols.Edinburgh 1801

Newton,I. Opticks,reprint of 4th edn.New York 1952;
Papers and Letters on Natural Philosophy, I.B.Cohen
(ed.),Cambridge 1958;
Unpublished Scientific Papers, A.R. and M.B.Hall(ed.)
Cambridge 1962

Nicholson,W. Dictionary of Chemistry,1795

Odling,W. A Manual of Chemistry,pt.1,1861;
Abstract of Lectures delivered at the Royal
Institution,1874

Oersted,H.C. The Soul in Nature, L. and J.B.Horner(tr.)
1852

Oken,L. Elements of Physiophilosophy, A.Tulk(tr.)1847

Ostwald,W. The Principles of Inorganic Chemistry,
A.Findlay(tr.),1914

Parkes,S. Chemical Catechism,1806; 10th edn.1822

Priestley,J. Disquisitions relating to Matter and
Spirit,1777;
The Doctrine of Phlogiston Established, Northumberland,
Penna.1800;
Scientific Autobiography, R.E.Schofield(ed.),Cambridge,
Mass.1966, includes bibliography of Priestley's
scientific writings

Prout,W. Bridgewater Treatise on Chemistry,Meteorology,
and the Functions of Digestion,2nd edn.1834

Ramsay,W. The Gases of the Atmosphere,1896;
A Life of Sir William Ramsay,by M.Travers,1956

Reid,T. Essays on the Intellectual Powers of Man,1785,
itr.B.Brody,Cambridge,Mass.1969

Rigg,R. Experimental Researches, shewing carbon to be
a compound body made by plants,1844

Roscoe,H.E. The Life and Experiences of Sir Henry
Enfield Roscoe,1906

The Collected Works of Count Rumford,S.C.Brown(ed.)
5 vols.Cambridge,Mass.1968-71

The Royal Institution Library of Science,Physical
Science,W.H.Bragg and G.Porter(ed.),11 vols.1970

Scheele,C.W. Chemical Essays,F.X.Schwediauer and
T.Beddoes(tr.),1786

Soddy,F. Radioactivity and Atomic Theory,T.J.Trenn(ed.)
1975

277

Somerville,M. On the Connexion of the Physical Sciences
1834; 9th edn.1858;
Molecular and Microscopic Science,2 vols.1869

Stallo,J.B. The Philosophy of Nature,Boston,Mass.1848;
The Concepts and Theories of Modern Physics,1888,
P.W.Bridgman(ed.),Cambridge,Mass.1960

Stevenson,W.P. The Composition of Hydrogen,and the
non-Decomposition of Water,2nd edn.1849

Stewart,D. Philosophical Essays,Edinburgh 1810

Tait,P.G. Recent Advances in Physical Science,3rd edn.
1885

Thomson,J.J. A Treatise on the Motion of Vortex Rings,
1883;
The conduction of Electricity through Gases,
Cambridge 1905;
The Corpuscular Theory of Matter,1907;
Recollections and Reflections,1936

Thomson,T. A System of Chemistry,4 vols.Edinburgh 1802;
3rd edn.1807; 7th edn.1831;
An Attempt to Establish the First Principles of
Chemistry by Experiment,2 vols.1825;
The History of Chemistry,2nd edn.2 vols.1830-1

Thomson,W. Notes of Lectures on Molecular Dynamics
and the Wave Theory of Light,Baltimore,Md.1884
(cyclostyled); printed edn.1904;
Popular Lectures and Addresses,3 vols.1889-94;
The Life of William Thomson,by S.P.Thompson,1910

Turner,E. Elements of Chemistry,3rd edn.1831

Tyndall,J. Fragments of Science,1871;10th imp.2 vols.
1899

Watson,R. Chemical Essays,3rd edn.5 vols.1784-7

Weldon,W. The Elements and General Laws of Chemistry,
1824

Whewell,W. History of the Inductive Sciences,3rd edn.
3 vols.1857;
William Whewell,by I.Todhunter,2 vols.1876

Wilson,G. Electricity and the Electric Telegraph,
together with the chemistry of the stars,new edn.1859;
Religio Chemici,1862;
Memoir of George Wilson,by J.A.Wilson,Edinburgh 1860

Wollaston,W. The Religion of Nature Delineated,1726
edition; the bibliography of this work is confusing

Young,T. A Course of Lectures on Natural Philosophy
and the Mechanical Arts,2 vols.1807

Secondary Sources

Benfey,O.T.(ed.),Classics in the Theory of Chemical Combination,New York 1963

Berman,M. Social Change and Scientific Organisation, London 1978

Boas,M. Robert Boyle and Seventeenth Century Chemistry, Cambridge 1958; as M.B.Hall,Robert Boyle on Natural Philosophy,Bloomington,Indiana 1965

Brock,W.H.(ed.),The Atomic Debates,Leicester 1967

Caldin,E.F. The Structure of Chemistry,1961

Cardwell,D.S.L.(ed.),John Dalton and the Progress of Science,Manchester 1968

Cartwright,F.F. The English Pioneers of Anaesthesia, Bristol 1952

Clagett,M.(ed.),Critical Problems in the History of Science,Madison,Wis.1959

Cohen,I.B. Franklin and Newton,Philadelphia,Penna. 1956;
Cohen,I.B. and Jones,H.M. Science before Darwin,1963

Collingwood,R.G. The Idea of Nature,Oxford 1945

Copinger,W.A. The Authorship of the first hundred numbers of the Edinburgh Review,Manchester 1895

Crombie,A.C.(ed.),Scientific Change,1964

Crosland,M.P. Historical Studies in the Language of Chemistry,1962;
The Society of Arcueil,1967;
(ed.) The Emergence of Science in Western Europe,1975

Dobbs,B.J. The Foundations of Newton's Alchemy, Cambridge 1975

Donovan,A.L. Philosophical Chemistry in the Scottish Enlightenment,Edinburgh 1975

Fox,R. The Caloric Theory of Gases,Oxford 1971

Fullmer,J.Z. Sir Humphry Davy's Published Works, Cambridge,Mass.1969

Fulton,J.F. A Bibliography of the honourable Robert Boyle,2nd edn.Oxford 1961

Gillispie,C.C. The Edge of Objectivity,Princeton 1960;
(editor-in-chief),The Dictionary of Scientific Biography,New York 1970-

Grabo,C.H. A Newton Among Poets,Chapel Hill,N.Carolina 1930

279

Gregory,J.C. The Scientific Achievements of Sir Humphry Davy,1940

Guerlach,H. Lavoisier — The Crucial Year,Ithaca,N.Y. 1961

Gunther,A.E. A Century of Zoology at the British Museum,1815-1914, 1975

Hall,A.R. The Scientific Revolution,2nd edn.1962

Hannaway,O. The Chemists and the Word,Baltimore 1975

Hartley,H. Humphry Davy,1966

Heilbron,J.L. H.G.J.Moseley,Berkeley and Los Angeles, 1974

Hesse,M. Forces and Fields, 1961

Houghton,W.E.(ed.),The Wellesley Index to Victorian Periodicals,1824-1900,2 vols.Toronto 1966-72

Ihde,A.L. The Development of Modern Chemistry,New York,1964

Jeffreys,A.E. Michael Faraday:a list of his lectures and published writings,1960

Kargon,R.H. Atomism in England from Hariot to Newton, Oxford 1966

Knight,D.M. Atoms and Elements,2nd edn.1970
Sources for the History of Science,1660-1914,1975
The Nature of Science,1977
Zoological Illustration,1977
Natural Science Books in English,1600-1914,1975
(ed.) Classical Scientific Papers,Chemistry, I,1968,
II,1970

Koyré,A. Newtonian Studies,1965

Leicester,H.M. and Klickstein,H.S. A Sourcebook in Chemistry,1400-1900,New York 1952

Levere,T.H. Affinity and Matter,Oxford 1971

Lindeboom,G.A. Herman Boerhaave,1968

Lovejoy,A.C. The Great Chain of Being,Cambridge,Mass. 1936

Lyons,H. The Royal Society,1660-1940,Cambridge 1944

Magie,W.F. A Sourcebook in Physics,New York 1935

McGucken,W. Nineteenth Century Spectroscopy,Baltimore, Md.1969

McKie,D. Antoine Lavoisier,1952

Meadows,A.J. Science and Controversy,1972

Mendelsohn,E. Heat and Life,Cambridge,Mass.1964

280

Merz,J.T. A History of European Thought in the Nineteenth Century,2nd edn.4 vols.1904

Meyer,E.von, A History of Chemistry,G.McGowan(tr.), 3rd edn.1906

Mottelay,P.F. A Bibliographical History of Electricity and Magnetism,1924

Nitske,W.R. The Life of William Conrad Röntgen,Tucson 1971

Palmer,W.G. A History of the Concept of Valency, Cambridge 1965

Partington,J.R. A History of Chemistry,vol.3,1962

Polanyi,M. Personal Knowledge,1958

Reingold,N. Science in Nineteenth-century America,1966

Ritterbush,P. Overtures to Biology,New Haven,Conn.1964

Rozental,S.(ed.) Niels Bohr,Amsterdam 1967

Russell,C.A. The History of Valency,Leicester 1971
Russell,C.A.,Coley,N.G., and Roberts,G.K. Chemists by Profession,London 1977

Schofield,R.E. Mechanism and Materialism,Princeton, 1970
The Lunar Society of Birmingham,Oxford 1963

Scott,W.L. The Conflict between Atomism and Conservation Theory,1644-1860,1970

Sinclair,B. Philadelphia's Philosopher Mechanics, Baltimore 1974

Singer,C.et al, A History of Technology,5 vols. Oxford 1954-8

Smyth,A.L. John Dalton,1766-1844,A bibliography of works by and about him,Manchester 1966

Spronsen,J.W.Van, The Periodic System of Chemical Elements:a history of the first hundred years,1969

Stafleu,F.A. Linnaeus and the Linneans,Utrecht 1971

Szabadvary,F. History of Analytical Chemistry, G.Svehla(tr.),Oxford 1966

Thackray,A. Atoms and Powers,Cambridge,Mass.1970; John Dalton,Cambridge,Mass.1972

Thorpe,T.E. Humphry Davy,Poet and Philosopher,1896

Wheeler,T.S. and Partington,J.R. The Life and Work of William Higgins,Chemist,Oxford 1960

Whittaker,E. A History of the Theories of Aether and

Electricity,2 vols.1951

Williams,L.P. Michael Faraday,1965

Williams,T.I.(ed.),A Biographical Dictionary of Scientists,1969

Whyte,L.L.(ed.),Roger Joseph Boscovich,1961

Index

Abernethy,J.,35

Accuracy, 39, 40, 163ff, 174ff, 178f, 185, 201, 218, 221ff

Acidity, 21, 63, 69, 70, 72, 74, 83, 127ff, 135ff, 189, 249

Aether, 45, 70, 74, 79, 80, 97, 100, 104, 106, 107, 108, 115, 117, 256, 258f

Affinity, 3, 4, 14, 15, 17, 20, 53, 103, 112, 128, 138, 225, 231, 248f

Air, 9, 18, 19, 34, 66, 73, 92, 158, 215

Alchemy, 9, 10, 13, 100, 114, 145, 149, 157, 198

Alkali metals, 23, 39, 53f, 63, 70f, 126, 131f, 147, 189, 190, 207f, 228, 247, 266

Allotropy, 190, 192, 233

Ampere, A.M., 163, 232, 237, 251ff, 254, 257

Analogy, iv, vii, 13, 20, 24, 35, 42, 61, 74, 79, 102, 117, 126, 137, 141, 143, 145ff, 155, 187, 190, 199, 204ff, 208f, 216ff, 222, 228ff, 239, 245f, 250ff, 258f

Analysis, v, 10f, 13, 17, 18, 19, 21f, 23f, 30, 41, 52, 66, 70, 129, 133, 156ff, 170ff, 178, 185, 191, 195, 208, 220, 225f, 253

Anderson,J., 237

Apparatus, iii, 10, 50ff, 63, 198

Aristotle, 9, 17, 18, 19, 20f, 24

Astronomy, 2, 14, 17, 143, 149, 166, 195, 205, 216, 224, 229ff, 262

Atoms, iff, vii, 4, 7, 10, 13, 15, 36, 63, 64, 67, 76, 82, 94, 98, 102, 105, 106, 113, 125, 137, 149, 158, 159ff, 169ff, 177, 186ff, 195ff, 198, 206ff, 218, 225ff, 231ff, 253ff, 262ff

Attraction, 15, 17, 20, 45, 69, 70, 81, 83, 97, 99, 102, 112, 114, 128, 231, 249

Averages, 164, 166, 263f

Avogadro,A., 109, 161, 163, 227

Babbage,C., 177

Banks,J., 32, 246

Beddoes,T., 34ff, 70, 97, 165

Bergman,T., 1, 3, 4, 34, 248

Berthollet,C.L., 4, 112, 128f, 138, 166, 216, 248

Berzelius,J.J., 4, 38,

Metaphysics, i, iii, v, 2, 5, 6, 13, 23, 29, 34, 61, 76, 84, 93, 125, 156ff, 162, 168, 200, 206, 268
Mill,J.S., 112
Mitscherlich,E., 188f
Molecules, 92f, 108, 160, 207ff
Morgan,W., 116
Moseley,H.G.J., 8, 258, 267

Naming, 37, 62, 101, 128, 138, 141, 162, 188f, 249, 251ff
National styles, v, 4, 5, 6, 38, 43f, 46, 63, 80, 108f, 126, 139, 146, 159, 173ff, 207, 216
Natural history, iv, 8, 23, 78, 80, 103, 144, 146, 159, 172, 202ff, 239, 246, 250ff, 257, 260, 265, 267
Natural theology, 2, 7, 61, 63ff, 68f, 74ff, 81ff, 96, 100, 105, 108, 144, 147
Nebulae, 107, 230, 236, 258
Newton,I., ii, iv, vi, 5, 7, 10, 14, 15, 33, 53, 54, 69, 73, 82, 99, 103, 105, 107, 114, 117, 146, 149, 195, 216, 239, 245, 247
Newtonianism, 2, 4, 7, 8, 17, 18, 33, 36, 61, 63, 70, 76, 80, 94, 96, 97, 98, 100f, 110, 116, 136, 161, 216, 245, 249
Nichol,J.P., 77
Nicholson,W., 32, 37, 40f, 46f
Nitrogen, 37, 38, 39, 61, 66, 71, 72, 100, 126, 131, 133f, 158f, 189, 193, 199, 207, 219, 230

Nominalism, 247f
Notation, 233ff, 249f

Odling,W., 29, 35, 233, 255
Oersted,H.C., 65, 114
Oken,L., 80, 206f
Order, 193, 245, 252
Organic chemistry, 165, 167, 186ff, 193f, 208, 225, 237
Ostwald,W., 238
Owen,R., 225
Oxygen, i, 18, 19, 21, 31, 38, 41, 44ff, 61, 63, 71, 72f, 83, 99, 100, 126ff, 137ff, 148, 158, 159, 193, 219, 221, 228, 232, 234ff, 250f, 255

Paracelsus, 9, 11, 17
Paradigms, iii, vi, 1, 6, 14, 21, 24, 36, 62, 91ff, 131ff, 143, 164ff, 203, 215, 246, 268
Parkes,S., 101, 155
Perspectives, iii, iv, vff, 61, 127, 163
Phlogiston, 4, 6, 16, 17, 24, 29, 44, 46, 63, 70, 72, 128, 132ff, 140, 149, 158
Physicists, viii, 14, 22, 62, 94, 169, 215, 235, 245, 258, 259
Playfair,L., 226
Polarity, 53, 65, 67, 109ff, 137, 196, 198, 206ff, 232f, 236
Popularity, 2, 40, 173, 238
Pourquoi non, 205
Powers, vi, 3, 4, 6, 15, 34, 52, 53, 61, 63, 67ff, 72, 73, 76, 78ff, 94, 106, 114, 115, 172
Prediction, 223, 224, 234, 236ff, 251, 256, 258, 259, 266
Priestley,J., 5, 18, 19f, 38, 44, 45, 51, 53, 64, 82, 112

Prime matter, 160f, 177, 207, 218ff, 222ff, 233, 236, 245, 256, 260ff
Principles, 9, 11, 12, 17, 22, 30, 35, 63, 65, 71f, 75, 100, 127ff, 133, 157, 185
Probability, 21f, 178, 215, 220, 239
Professionalism, iii, 1, 39, 173, 203
Proof, 21, 136, 141, 158, 171, 221, 228, 245f, 256
Properties, 63, 76, 77, 95, 100, 106, 113, 115, 125ff, 133, 148, 165, 186, 194, 205, 215, 248ff, 254ff
Proportions, 127, 148, 157, 162, 171, 187ff
Prout,W., vii, 77, 108ff, 126, 144, 149, 155, 159ff, 165, 169ff, 177f, 185, 196, 215, 217ff, 239, 254, 256f
Purity, 10, 47, 50f, 129ff, 136, 170ff, 175, 239
Putrefaction, 201ff

Quantification, vii, 4, 31, 42, 46, 52, 94, 104, 112, 114, 126, 128, 155, 157, 159ff, 169ff, 176f, 186, 199, 218ff, 227f, 234ff, 239, 248ff, 253ff

Radicals, vii, 30, 61, 72, 94, 147, 187, 192ff, 207, 217ff, 225, 239, 250, 254f
Ramsay,W., 8, 200, 260 266
Ratios, 161ff, 176, 207, 219, 222ff
Rayleigh,Lord(Strutt,J.W.) 239
Reid,T., 21, 82
Repeating Experiments, 175, 198ff, 221
Revolution, 14, 39, 139
Reynolds,E., 259
Rigg,R., 201ff, 209

Ritter,J.W., 38, 43, 45, 80, 96, 102, 149
Romantics, 7, 77, 79, 95
Roscoe,H., 237f
Royal Institution, iii, 44, 48, 68, 96, 100, 103, 142, 146, 171
Rutherford,E., ii, viii, 8, 157, 266f

Scheele,C.W., 18, 19, 34, 128, 138
Schleiden,M.J., 204
Scientific societies, iii, 2, 23, 34, 37, 44, 48, 49, 92, 96, 97, 100, 105, 142, 145, 173, 189, 201, 216f, 225ff, 231, 233, 237, 260, 261
Semantic change, 62, 68
Simplicity, ii, vi, 5, 8, 10, 16, 19, 22, 29, 34, 48, 61, 108, 114, 127, 134, 137, 142ff, 147, 149, 156, 158, 159ff, 165ff, 169ff, 187, 192ff, 195, 206, 209, 215, 225, 228, 235ff, 245f, 262, 268
Soddy,F., 8, 266
Spectra, 8, 24, 61, 103, 107, 229ff, 236, 239, 256, 266
Spirit, v, 5, 64f, 73, 74, 76, 81ff, 96, 106, 107
Stahl,G.E., 6, 13, 16, 17
Stallo,J.B., 216
Stas,J.S., 217, 220ff, 239
States, 6, 8, 68, 92ff, 99, 104, 106ff, 125
Stewart,D., 156f
Stokes,G.G., 234f, 260
Structure, vii, 4, 12, 14, 24, 36, 53, 73, 74, 75, 100, 110, 113, 128, 137, 148, 169, 186ff, 194, 225, 234ff, 255
Subtle fluids, 62, 65, 70, 76, 94, 104, 115, 116, 129
Synthesis, 6, 13, 18, 24,

75989   QD
                                18
                                G7

KNIGHT, DAVID
~~~ mbc mpaucaennemat dapm om~~~

DATE DUE	
GAYLORD	PRINTED IN U.S.A.